ENVIRONMENTAL RESEARCH ADVANCES

AN IDEAL ECOSYSTEM AND SEVERAL PROBLEMS OF OUR TIME

Environmental Research Advances

Additional books in this series can be found on Nova's website under the Series tab.

Additional e-books in this series can be found on Nova's website under the e-book tab.

ENVIRONMENTAL RESEARCH ADVANCES

AN IDEAL ECOSYSTEM AND SEVERAL PROBLEMS OF OUR TIME

YURY KAMENIR
MARINE ECOLOGY LAB., BAR ILAN UNIV.,
RAMAT-GAN, ISRAEL

nova publishers
New York

Library of Congress Cataloging-in-Publication Data

ISBN: 978-1-63117-300-4

Published by Nova Science Publishers, Inc. † New York

CONTENTS

PREFACE

The purpose of this book was to discuss some general results of studies devoted to the structure and self-maintenance of natural aquatic ecosystems, as well as concepts attempting to explain the observed phenomena.

Recently, we begin to note that something helps us live. Such a helper has already earned a special name full of our gratitude – ecosystem (from the Greek word oikos, meaning home). Numerous processes sustain or enhance human life. However, we often pay attention only when some unusual inconveniences arise. Such troubles have recently become more frequent; hence, we feel an ever-growing need for sound knowledge of our sweet home.

Our world is changing. Undesirable changes are accelerated by our deeds and faults, and cannot be ignored. We should not be lazy or stupid. The global warming parallel to the industrial revolution and the dramatic growth of human populations might be a most plausible explanation for the crisis origin; hence, human activity (i.e., our own scientific-technological revolution) is to blame. The rapid evolution of human intelligence is not only the cause but also the most realistic instrument for finding solutions for the tremendous problems of our time.

Rapid progress in biology is evident, but much more is needed. Many kinds of ecological diagnoses and forecasts seem to be far from solution. Impressive advances are evident in ecology and theoretical biology, especially, in genetics, biomedical statistics, mathematical modeling and informatics; such sciences efficiently support theoretical explanations of typical patterns of living bodies and assemblages. Allometric scaling of numerous traits of living organisms can now be reasonably explained with the help of very general quantitative models depicting the transport of material and energy through space-filling fractal networks.

A special field of science, i.e., the art of modeling has been developed. Abundant tools and methods were developed supporting modeling of living systems, e.g., the general theory of systems, various mathematical approaches and quantitative generalizations independent of specific mechanisms in living objects. Several optimization principles were formulated, including the Ockham's lex parsimoniae, i.e., the law of economy of thought. The number of parts, links, and traits described should be a minimum. Such models are called minimal; we can see the minimal cell, the minimal genome, and the minimal ecosystem models. For analyses of environmental problems, minimal models of living communities and ecological systems are needed. Therefore, we started a dramatic journey into forests of literature and empirical data taking into consideration living assemblages that support our existence.

Long years of friendship and collective work have brought about countless pleasant moments (finding interesting facts, ideas or solutions to small problems). Many moments were sad - the results and ideas found seemed to be too unusual; hence, much time and troubles were spent publishing and discussing the findings and ideas. Several colleagues have already finished their work and life, but others continued the same route.

Oleg Makarov was the first professional mathematician to provide me with support and administrative assistance during several difficult years when ideas of living matter seemed strange to people influencing our work plans. Thirty-forty years ago, many specialists believed that the fantastic progress of computer capacities (first of all, the calculation speed and the memory storage) would solve all ecological problems. Now we have tremendously quicker computers capable of handling models with hundreds of parameters and millions of equations. Computer-made simulations have become the everyday games of teenagers, but the basic mechanisms supporting ecological systems and living assemblages are far from being understood. A special way of thinking is needed to progress from the fabrication of iron-made mechanisms to the analyses of multitudes of living cells and assemblages.

Kirill Khailov, a brilliant enthusiast of ecological metabolism had given me lessons and ideas on Russian cosmism, especially, the living matter, geobiochemistry, the biosphere and the noosphere - concepts pioneered by V.I. Vernadsky. Timur Ayzatullin supported my interest in ideal ecosystem modeling. Victor Gorshkov moved from nuclear physics to environmental-protection mathematics, gave advice on allometry and size distributions that have recently been given a special 'physics-like' name - size spectra. Danara Krupatkina and Alexander Lopukhin supported my early ideas with empirical studies in the Tropical Atlantic and Indian Oceans. Sergey Kovardakov and Alexander Prazukin provided empirical data on the size structure of hierarchical structures of large marine plants (macrophytobenthos). All together they have given several years of work to metabolic-surface estimation and analysis of the variability of dimensionless (m^2 m^{-2}) indices of the metabolic surface of living assemblages.

The fundamental collections of biodiversity distributions of Leonid Chislenko and the energetic support of Alexey Armand, Georgy Winberg, Michael Vinogradov, Tatiana Bek, and Victor Maximov, has created a bridge to readers and scientific discussions. Numerous unknown people provided support needed to see unique ecosystems of pristine Lake Baikal and the White Sea, the Kamchatka Peninsula, fantastic anthropogenic destruction of the Aral Sea, and anthropogenic transformations of the marvelous Solovezky Archipelago SLON, which gave much information on the limit of ecological variability and the two-sided influence of human creativity. Further meetings with foreign colleagues studying size-spectrum patterns – Susana Agusti, Carlos Duarte, Kent Denman, Garry Sprules, Jaime Rodriguez, Jeffrey Nekola, and Jordan Okie – provided fruitful discussions on size spectra.

Thanks to a valuable opportunity to scan old French scientific books in Paris, I have found the living whirl ideas of G. Cuvier. This intellectual support was very important to the continuation of our studies on living matter. Several rounds of comparative studies brought additional hints as to the formalization of general patterns of ataxonomic size spectra. Boris Shteinman, Tamar Zohary, Moshe Gophen, Paul Walline, Talia Bergstein Ben-Dan, Tom Berman, Baruch Kimor, and Zvy Dubinsky gave me invaluable data and literature on Lake Kinneret (one of the best examples of multi-annual stability of phytoplankton assemblage) and the history of its drastic change. These ideas and data gave us the cornerstones of taxonomic size spectra. Bibliographic scanning has provided data on several long-term studies of artificial systems. They were added for purposes of comparison and help us expand the ecosystem size range applied to develop regressions considering the ecosystem-size influence on living-matter estimates and largest organism body size of aquatic ecosystems.

Richard Kemp, Tatiana Mikhailyuk, Antonina Popova, Monika Winder, Geoffrey Schladow, and Giuseppe Morabito have demonstrated the intricate differences between several lacustrine ecosystems and the long-term consistency of general patterns of taxonomic size structure of phytoplankton assemblages. Even more significant were comparisons of marine phytoplankton with freshwater phytoplankton, developed with Roger Harris and the highly qualified staff of the Plymouth Marine Laboratory. Marvelous work (multi-annual monitoring) was accomplished by people studying Lakes Kinneret, Tahoe, and Maggiore. Large quantities of data created opportunities to develop a more sensitive modification of taxonomic spectra, i.e., frequency-weighted size spectra that amplified our diagnostic capabilities; we could analyze the size-structure consistency of major phytoplankton phyla. The following addition of analyses of biodiversity histograms and rank distributions has led to the multisided analyses of aquatic living matter.

With the help of large-scale empirical works, good fortune, and specially selected datasets, we could find some traits and general patterns of living matter of highly different ecosystems, from artificial microcosms to the whole world ocean. After several decades of studies, the ideas, methods, and models depicting the ecosystem stability phenomenon look a little different; however, the most general themes (living matter, living whirl, ideal ecosystems) and general thoughts on the ecological crisis are the same. Our joint property – the biosphere – our home is very strong. We can survive even if some of us are not sufficiently clever; however, we must be careful.

Most of material here was written for specialists interested in the modeling of biological systems and environment protection, and for ordinary people interested in the ‘constructions’ of the living systems that surround us and support our existence. Numerous special terms and mathematical methods are explained, and scientific publications have been added for better understanding, and can be found in the bibliography referenced here.

It is my great pleasure to thank the numerous co-authors and colleagues who provided empirical material and theoretical considerations. Experimental data for cooperative works were collected at several institutes – the IBSS (USSR), the Bar-Ilan University, Haifa University, and KLL-IOLR (Israel), the Lake Taho Laboratory (USA), the Institute for Ecosystem Studies (Italy), and the Plymouth Marine Laboratory (Great Britain).

Numerous people unknown to me from Nova Science Publishers supported several of my publications on size spectra. I deeply appreciate the advice and support of Sharon Victor for typescript editing.

Yury G. Kamenir
Research Fellow
The Mina & Everard Goodman Faculty of Life Sciences
Bar-Ilan University
Ramat-Gan 5290002, Israel
kamenir@biu.ac.il

November 11, 2015

Chapter 1

OUR WORLD IS CHANGING

Our milieus (i.e., our home and surroundings) provide countless services that we generally take for granted. Recently, more and more often we begin to note that something helps us live. Such surroundings have already deserved a special name full of gratitude – ecosystem (from the Greek word *oikos*, meaning home). Numerous processes (the ecosystem services) sustain or enhance human life. Many benefits are derived from well-functioning ecosystems (Brown et al. 2007). However, we often pay attention only when some unusual inconveniences arise. Recently, such troubles have become more frequent; hence, we feel an ever-growing need for sound knowledge or, at least, the more critically important details of the design of our sweet home (Whittaker & Likens 1975).

Thanks to long-time experience and numerous publications, we can note the tremendous complicacy of life (Odum 1971). A great amount of information is needed to describe and imagine even the smallest living organism. No less challenging are problems facing each researcher who studies living assemblages, e.g., a forest, a lake inhabited by various animals and algae, or even a drop of water with a myriad of living particles moving under the microscope (Sieburth et al. 1978). Due to the remarkable progress of modern science, especially computers, molecular biology, and automated information-acquiring tools, we can accumulate huge data stores for the development of theoretical schemes and environmental monitoring. Such databases provide the basis for ecological protection and forecasting; however, such forecasting is far from accurate, due to numerous strong impact factors (Huisman & Weissing 2001).

Analysis of the structures and functioning of living systems is often rather complicated; interpretations are a lengthy process, much slower than what is demanded by practical needs. Solutions are urgently needed and delays are intolerable; however, now, we do not have just an impressive collection of troubles but an integral huge problem, a real global environmental crisis (White 1967; Vitousek 1994). Our world is changing at an unprecedented rate. While undesirable changes are accelerated by our deeds and faults, they cannot be ignored. We should not be lazy or stupid. We must adapt or perish (Sagan 1977). Huge areas of the globe are now unsuitable for humans or problematic to our survival, e.g., good pastures are transformed to deserts, agricultural space is consumed by growing cities and factories, lake depths are diminishing, and huge seas are disappearing from the globe. Other vast regions (e.g., in Holland or Bangladesh) might be covered by water if the world ocean level continues to rise. The global warming parallel to the industrial revolution and the dramatic growth of human populations (Arrhenius 1896; Vitousek 1994; Shakun et al. 2012), might be a most plausible explanation for the crisis origin; hence, human activity (i.e., our own scientific-technological revolution) is to blame. The rapid evolution of human intelligence is not only the cause but also the most realistic instrument for finding solutions for the tremendous problems of our time (Sagan 1977).

Rapid scientific progress is currently being made, but much more is needed. Many kinds of ecological diagnoses and forecasts seem to be rather far from solution. Tremendous progress is evident in some fields of biology, e.g., molecular biology, genetics, global environmental monitoring, etc. Impressive advances are evident in ecology and theoretical biology, including mathematical biology and informatics, supporting huge empirical data sets and theoretical explanations of typical patterns of living bodies and assemblages (Peters 1983; Sprules & Munawar 1986). Allometry, also called biological scaling, describes the effect of living-body size on numerous quantitative properties of living organisms (Peters 1983). Many fundamental traits of living organisms scale with body size in a simple, general way. Universal theoretical models were developed that are capable of explaining how many living systems are built, supported by hierarchical branching networks. Such space-filling networks produce metabolic flows supporting various living structures (West & Brown 2005).

Some basic allometries taken together produce invariant quantities independent of organism size. For example, the number of heartbeats per lifetime is approximately invariant (~1.5 billion) for organisms of very different origins and sizes (Charnov 1993). Such invariants support some construction rules selected and safeguarded by the evolutionary process. Some general principles have been widely applied in ecology and evolutionary theory to predict the attributes of living organisms (Charnov 1993). Among general ecological regularities is size spectrum (SS), i.e., a range of observed objects (in our case, a range of living organisms or other different-sized structures). Two most prominent types of size spectra (BSS and NBS) describe rather stable 'typical' patterns of biomass distribution fulfilled by assemblages of living organisms (Sprules & Munawar 1986; Kerr & Dickie 2001). Combinations of such cornerstone mathematical tools can produce more complex, multi-sided models and theories.

Sometimes models are too large and complicated for an individual human brain. Computers can help with many kinds of modeling problems. A challenging process is computer model development and minimization. Parsimony (i.e., a principle of economy or succinctness in the application of elements and connections) is a very challenging gauge leading to ecological model evolution and theoretical biology development (Sober 1983). Starting from extensive textual descriptions and elaborate details, we proceed to more and more general and less complicated mind constructions suitable for practical exploitation, teaching, and further generalization. Recent history provides numerous examples of analyses transforming multi-volume treatises of specific biological problems into very general and laconic rules that every schoolboy takes for granted without considering tremendous setbacks and doubts that took long years to overcome. A very efficient modeling tool is information compression, i.e., the transformation of huge stores of loosely connected fragments into a compact model or a concept leading to an efficient solution. A compact, intensely idealized model of the object of our studies is needed, just as modern physics has developed the ideal gas notion (Levine 2002), the planetary atom model of Bohr, and the very compact Newton's set of mechanics laws (Tipler & Llewellyn 2002). In more distant history, we can mention the change in astronomical models, with one cosmic body (our Mother Earth) taking the center of the world, and later giving it to the sun (Hoskin 1999). Models of two-dimensional and 3D geometry have served man for many centuries, but are lately overcome by more general, non-Euclidean models of geometry (Greenberg 1993). Allometric scaling of metabolic rates, characteristic of all organisms, can now be reasonably explained with the help of very general quantitative models describing the transport of material and energy-carrier substances through space-filling fractal networks (West et al. 1997). Similarly, for analyses of environmental problems, a minimal model of ecological systems seems to be an efficient start, a focal point for more elaborate studies. A reasonable step in this direction can be called the living whirl.

Chapter 2

LIVING MATTER

INTRODUCTION

Our aim is to search for simple (compact and easy-for-perception) models describing our living home – the Earth – and its ecosystems. Looking at our dwelling place and nursery, we understand that its principal component is a huge ensemble of self-regulating living communities protecting us and supporting our existence within closed walls. Numerous definitions of an ecosystem (e.g., Begon et al. 1996) cite a set of populations of living organisms and a set of abiotic factors of the environment. A typical definition is given by the world-level popular Wikipedia: An ecosystem is a community of living organisms in conjunction with the nonliving components of their environment, interacting as a system. These biotic and abiotic components are regarded as being linked together through nutrient cycles and energy flows (http://en.wikipedia.org/wiki/Ecosystem, entered 21.04.15). Several principal components head the list of the minimal set of traits necessary to describe ecosystem existence: living organisms, cyclic movement of matter [material components (including specific chemical elements) and nutrients], and the energy supporting such movement. The subsequent choice of alternatives is to analyze a vast list of components (e.g., chemical elements, molecular complexes, and biological species) or start from one (the most essential) property of the whole object. Going to just one focal point, we can add further long chains of details - components, forces, and traits – providing, step by step, with ever growing precision and completeness, the properties of the whole object. The subsequent polishing goes from the top – down, from one specific principal point.

Such an approach, i.e., the top-down design (also known as stepwise refinement and often used as a synonym of decomposition) helps us break down a system into a hierarchic set of elements, links, and properties to gain understanding of its essential traits. By the top-down approach, an overview of the system is formulated, specifying its first-level subsystems. Then, several rounds of refinement describe the whole system in yet greater detail, sometimes at many levels of its subsystems. Step-wise refinement is applied as a powerful paradigm for developing complex research programs by adding features of the studied object incrementally (Batory et al. 2004).

Thus, one word – ecosystem (from the ancient Greek word "οἶκος," meaning 'home') introduces the definition. The main feature of such an object is a well-protected space or area suited for the survival of a living being. Consequently, an ecosystem is not just any space, but rather an inhabited space or one suitable for habitation; it is a place suitable for a living body or an ensemble of living entities. Therefore, this space is characterized by a set of traits suitable for life (for maintaining life, for long-term habitation). We can define the habitable zone in terms of the chemical and physical conditions required for life (Pace 2001).

Life on our planet is supported by liquid water around, and especially within, numerous living particles (organisms); consequently, it exists within a small range of temperatures (some 100ºC). Some principal components of living particles (e.g., proteins) are destroyed by warming; hence, the temperature range suitable for life is even smaller. There are many organisms dwelling in the air (flying in the Earth's atmosphere or on the planet surface) and even inside terrestrial soil; however, all of them exploit some water sources to replenish their inner liquids.

All living particles (organisms and colonies) are composed of one or several cells (Begon et al. 1996). Each such particle has limited longevity (a time limit till its death); however, some particles produce new particles resembling their precursors. The living cell is the smallest unit of life that can replicate independently. All cells arise from preexisting cells through cell reproduction (Farabee 2006).

Each cell consists of several layers. The inner layer - protoplasm - is enclosed by a membrane and contains the living contents of the cell. Protoplasm is composed of water and a mixture of small molecules (e.g., amino acids) and macromolecules, such as proteins and nucleic acids. Nucleic acids (DNA, RNA) and proteins enable cell replication and development (ontogenesis) (Farabee 2006).

The outer layer (cell membrane) is critical for cell life and death. Cell viability can be checked by a membrane permeability probe (Perfetto et al. 2006). The cell membrane is a very remarkable closed envelope with a set of interesting properties. Its most evident basic property is that it protects the cell. Just as each cell has such a protective envelope, each organism (one or several living cells) is encircled by a specific surrounding shell, including the Earth's atmosphere (gases, dust, vapor) and world oceans (liquid water). This liquid water is seen as planetary oceans, seas, lakes, and (the especially important part!) the water within all living cells (Lebedev et al. 1989). If the water inside a cell is frozen or its concentration too small (e.g., in grains, nuts, etc.), cell development is suppressed. Special molecular mechanisms help some cells (e.g., Archaea) survive within a specific temperature range (approximately 0 to 100ºC), though the main part of living species and biomass can survive only within smaller temperature ranges (Koga 2012).

To have such an envelope of water and gases, our home should be a planet, i.e., a globe of definite mass, chemical composition, and space conditions, e.g., situated near a star (a source of light, our Sun), at a specific distance, providing the solar energy and temperature regime of the planet. Many planets do not have such an atmosphere; however, modern astronomy gives evidence of the existence of numerous planets that may have such an envelope (i.e., atmosphere and liquid water) (Seager 2013). This envelope works as a specific membrane or filter, permitting the entrance of some radiation from outer space (the solar light) while preventing entrance of many other types of radiation (high-energy radiation can damage or even kill living organisms). Again, the cell membrane parallels the Earth's envelope: each cell membrane works as a specific filter, preventing many kinds of material, energy, and information flows, whereas some flows are permitted. Such a filter is an indispensable part of any living system. Many kinds of chemical stuff are dangerous to living organisms, but some of them are beneficial and even indispensable for living organisms

(Odum 1974; Begon et al. 1996). Although some chemical components are beneficial, they can be dangerous if their concentrations or ratio between elements (the stoichiometry) change. Chemical composition of the environment (i.e., the living and non-living matter around a living particle) is very important for survival of such organisms. There are many types of living organisms; some of them use or produce chemical stuffs dangerous to other living forms. Therefore, the inhabited space is structured, and divided into compartments with different fillings, e.g., the ocean, freshwater lakes, swamps, atmosphere, various soils, etc. (Odum 1974; Begon et al. 1996).

Just like a living cell, each ecosystem has a store of water, living matter, and non-living matter. Just like a living cell, each ecosystem has limited longevity. The ecosystem longevity depends on the activity of its filling, i.e., its living matter (LIM) and non-living matter (bio-inert matter or BIM) created or processed by LIM. Some BIM varieties create envelopes that shield the ecosystem, e.g., active films (layers) at border zones between the water body, atmosphere, and soil. At such border zones, concentrations of numerous components of LIM and BIM are considerably elevated (Lebedev et al. 1989), similar to the shell and flesh of a living organism.

LIM is composed of multitudes of particles (cells) that have several general properties: metabolism (i.e., transformation of material, energy, and information), reaction to external stimuli, homeostasis (i.e., specific feedback properties preventing changes in some critical traits of the cell), growth and fission, death and progeny reproduction, etc. (Farabee 2006).

The most interesting property, called ontogeny, integrates several principal properties of a living system. The ontogeny or ontogenetic cycle is a complex self-maintenance process of the living system that is characterized by the replacement of ageing subsystems by younger and less-damaged copies. Hence, it can be seen as a specific variant of homeostasis, where particles predestined to pass away (e.g., to die) are thrown out (i.e., they perish, are broken down, dissolved, etc.) and replaced by younger particles (better suited to current conditions) produced via a long developmental process completed under suitable time limits (i.e., the living particle durability). Ontogeny is a complex process by which the hereditary information contained in each cell unfolds at several stages of development, starting from living-system reproduction (e.g., cell fission), continuing with the smaller cell growth, maturing, new progeny preparation, and finishing with the 'younger' (now, a little older) organism – death (Begon et al. 1996).

Reproduction is a fundamental feature of all known life, one of main criteria needed to describe living systems. All known types of organisms are capable of reproduction. Some important organic structures (e.g., viruses) are incapable to independent reproduction. Several types of reproduction processes are known (cell fission, asexual or sexual reproduction). The most interesting (complicated and important for us) type of reproduction process goes via formation of a zygote, that is application of one reproductive cell transformed by a special procedure – fusion with a sex-partner cell - to generate a new combination (recombination) of genetic material from those two reproductive cells. Genetic recombination is the production of offspring with combinations of traits, dissimilar to any parent. A zygote (one cell) begins a new ontogenetic cycle creating omnipotent stem cells progressing into a living organism, a much larger multicellular system composed of several organs formed of several tissues of numerous cells. Totipotent or omnipotent stem cells can differentiate into numerous cell types. Subsequently, omnipotent cells can construct a complete, viable, highly heterogeneous system (a multicellular organism). Potency of a cell specifies its potential to differentiate into different cell types (pluripotent, multipotent, oligopotent, unipotent). Unipotent cells can produce only one cell type. Thanks to recombination and spontaneous hereditary information changes (mutations) huge variability of organisms and specific cells was formed. Multicellular organisms can contain a wide range of different cells, some of them being reproductive cells. A large organism can produce reproductive cells that become zygotes of new organisms. So, the ontogenetic cycle is the key process of LIM temporal structure. Many living particles (organisms, cells) do not produce any offspring. While any organism has a limited durability, some of organisms and cells succeed to produce the offspring, and myriads of living cells continue the eternal process of life on our planet (Begon et al. 1996; Farabee 2006).

We shall call particles of LIM 'flow-through elements' (FEL). A very interesting general property of FELs is that some of them consist of smaller FELs (with respective envelopes, e.g., cell organelles, cells of a multicellular organism, organisms of a population, populations of an ecosystem, etc.); hence, we see hierarchical structures. Sets of walls protect our house, just as a membrane encircles and protects each living cell. A specific membrane layer envelopes each living organism, e.g., our skin, shell, bark wood. Such high-gradient layers are notable at each marine shore, the sea bottom, air-water interface, and at the air-soil boundary (Lebedev et al. 1989; Strayer et al. 2003).

Again, LIM consists of multitudes of discrete particles. The flow of mobs of ephemeral (mortal) entities, plus the activity of such mobs, provides a feedback stabilization of their common milieu. The FEL-habitat traits should be maintained at least for a definite time interval critically important for living-body ontogenetic cycle to be finished. The ecosystem traits should suit the inhabitant. What is suitable for a bacterium may be unacceptable for a whale or another large creature; even more so, it may be intolerable for a huge mob of small creatures or an ensemble of various creatures.

Speaking about an ideal object, we can disregard many obstacles (fluctuations or instabilities, rhythmicity, variability, intrinsic noise, etc.) while trying to stress the most relevant traits. One of most important properties of each FEL (including a cell, organism, and ecosystem) is its longevity. The ideal ecosystem would have the ideal ability to exist, i.e., it would exist eternally. No real object has such properties, but some objects are more durable than others. According to numerous treatises, LIM is a strange alternative to other longevity record-breakers - extraterrestrial structures existing at very low thermodynamic activity (Glansdorff & Prigogine 1971). Other remarkable objects are rocks and crystals born under extremely high temperatures and under huge pressure of planetary depths, to exist unharmed on the planet surface, i.e., within 'our ordinary environment.' LIM inhabits environments situated between such two extreme conditions. Living objects are concentrated on the planet surface; hence, they do not go into the hot central zone of the Earth nor to 0°K zones typical of interstellar structures. Living structures inhabit much warmer milieu; they, themselves (the living flesh), are much warmer (some 300°K); they always transform and dissipate energy (Glansdorff & Prigogine 1971) and, consequently, they maintain a much warmer environment. The dead body breaks down quickly (due to activity of the living microorganisms within or around such a body); hence, special measures (e.g., strong cold) are necessary to prevent the destruction of living structures. Dissipative structures and the hierarchy of cycling flows are the cornerstones supporting a minimal model of LIM (Kamenir 1983, 1986; Kamenir & Shteinman 1998).

Chapter 3

LIVING VORTEX

INTRODUCTION

The thin shell of the Earth, filled with the living objects, has a number of marvelous properties. First of all, LIM exists and can be felt almost everywhere. It is not only an abstraction seen from scientific treatises or poetic verses. It is a kind of matter, a set of material (visible, measurable, verifiable, and suitable to study) objects. One cannot escape living objects around you, and even inside your own body. Sometimes we see living organisms, sometimes we feel their presence, sometimes we are not aware of their existence and work, but we consume fruits of this work. We take it for granted, like drinking water, milk, honey, and the fresh air we breathe.

Secondly, LIM is everywhere; there are no empty holes. You should make a huge effort to find a location where there are no living objects (Sieburth et al. 1978). Even at the coldest polar regions or the hottest 'lifeless' deserts, there are numerous living organisms (often they are invisible microorganisms). They are very different from us, but can now be found and visualized by means of modern scientific tools. The third geometrical coordination of our planet (i.e., the vertical profile) shows us that LIM is concentrated at the planet surface; its concentration diminishes many orders of magnitude at a tiny distance (several km, compared to the Earth's diameter) going upward to the atmosphere or downward into the soil. The largest vertical dimension of the LIM shell is in deep seas and oceans, i.e., just several kilometers. Therefore, one can describe the LIM space configuration as a thin film or a planetary skin (Vernadsky 1978; Lebedev et al. 1989). Just like our own skin, this thin layer is very resistant, and efficiently protects the Earth.

This thin layer is highly active, yielding a remarkably great amount of chemical transformations. All LIM particles are interconnected through the biogenic migration of atoms via respiration, organic-matter production, excretion, grazing, parasitism, child production, etc. As a rule, the biogenic migration of atoms involves not only other living particles, but also non-living particles, liquids, and gases. LIM is among the most active and efficient transformers of material, energy, and information stores. LIM is very active, both chemically and geologically, and carries out several important functions. Myriads of LIM particles infiltrate huge spaces of the biosphere, acquiring and expending huge amounts of energy. This energy provides a great amount of physical and chemical work (moving, crashing, eating, and dispersing both huge blocks and microscopic particles of inert, bio-inert, and organic matter, synthesizing, destroying, dissolving, and excreting). LIM constitutes a great geological force, changing the face of our planet and the internal structure of its surface layers. The geological role of such transformations is seen in sedimentary structures (e.g., coal, bitumen, and petroleum) (Vernadsky 1978). The atmosphere, soil, and almost all the water of our planet are produced and continuously conditioned by LIM. Such bio-inert materials form the surface layers of our planet and surround us. They are not always visible, but critically needed.

The biogeochemical functions of LIM are carried out both inside living organisms and in the external medium. Huge masses of gases on our planet are biogenic; the gaseous composition of the atmosphere is transformed by living matter. Especially important is the LIM role in N, O, C, H, S, and some other elements. The concentration functions of LIM include the accumulation of biogenic (i.e., essential for fundamental life processes) elements from the surrounding medium. The chemical composition of LIM is clearly different from that of inert matter.

LIVING VORTEX

A good operational definition of life is a self-replicating, evolving system (Pace 2001). The basic drive of life is to make more and more of itself. Self-replication and genetic continuity are fundamental traits of the life process. Complex molecules support all processes of life known on our planet. Such molecules are made up of carbon bonded with other light elements, especially oxygen, hydrogen and nitrogen (C-O-H-N). Carbon has four valence bonds and the energy level appropriate to build and modify long complex molecules. There are not many alternatives for such functions. Carbon is one of the most abundant chemical elements. Therefore, carbon-based molecules are probably rather common throughout the universe. Life also must capture energy and apply that energy for replications. The electronic properties of carbon allow the formation of double or even triple bonds. The molecules that serve living organisms typically are very large - proteins, DNA and RNA - the building blocks of living organisms, genetic information maintenance, plus carbohydrates and lipids for energy stores (Pace 2001).

The LIM particles contain predominantly light atoms (H, C, N, O), but important are also several other elements (e.g., P, S, K, Cl, Si, Ca). Chemists apply a special term, biophilic (i.e., life loving), to describe these elements. Some metals play key roles in biochemical processes, in very small quantities. In larger concentrations such elements are very dangerous for living cells and are considered as pollutants. The concentrations of certain chemical elements are very different for the Earth's crust, rivers, oceanic water, and living matter. The concentrations of certain elements in some organisms are many orders of magnitude higher than in the external medium (Korzh 1985; Lebedev et al. 1989). Such LIM activity results in heterogeneous (very dissimilar) chemical composition of the biosphere; both the living and bio-inert parts of the Earth's shell are highly diverse.

One of most important functions of chemical transformation is called the oxidation-reduction of LIM. Oxidation-reduction (or redox) reactions involve the transfer of electrons between chemical species. Such reactions are common to fire (combustion, dissolution of metals), corrosion, and also to the basic functions of life (photosynthesis and respiration).

Biogenic oxidation and reduction predominate in the surficial layer of our planet. This continuous process transforms most substances. LIM is constructed from a limited choice of chemical elements (see the Periodic Table). The reduction of oxygen provides the largest free energy release per electron transfer. An atmosphere rich in oxygen and oxygen stored within the world ocean provide a huge energy source. Abundant oxygen is necessary for the high-energy demands of complex LIM, for mobile organisms, and for large organisms capable of efficiently processing information (Catling et al. 2005). Therefore, processes involving oxygen (oxygenic photosynthesis and respiration) are of critical importance for our planet and its inhabitants (and have been for billions of years!) (Odum 1971; Catling et al. 2005). Liquid water (the world ocean) and the oxygen-rich atmosphere are two critical layers of our planet. The biosphere and biogeosphere represent the layers of the Earth where life can exist. That is the ocean, soils and sediments (i.e., parts of the crust), and the lower layers of atmosphere. The upper solid layer of the Earth comprises some 1.2%, plus the atmosphere and ocean together $< 0.025\%$ of Earth's mass. These layers support particles of LIM composed mainly of three chemical elements, i.e., O, H, and C (Schaub & Turek 2011; Milo & Phillips 2015).

The biochemical functions are connected with the growth and multiplication of organisms that are always hungry for matter, energy, and information. The 'aggressive intrusion' of LIM into the ambient medium was called the pressure of life by Vernadsky (1978). This pressure can be seen from the biomass production, growth, and energy stocks stored as living and dead organic mass. This function is very variable as we consider different groups of organisms, but all of them must take in and eject material components from the surrounding medium.

The biochemical function of living matter also includes the decomposition of dead organisms, transforming LIM into inert matter. The biogeochemical functions of mankind (i.e., just one of a multitude of species inhabiting the Earth) are huge in number, and include a huge variety of forms. Such changes, which are produced by man (i.e., anthropogenic), are evident in any ecosystem and in the whole biosphere. Look, for example, at global deforestation, agricultural development, and urbanization, all created by humans. Such large-scale transformations are seen from satellites and from the Moon. They are tremendously accelerated by human technological activity (i.e., technogenesis). The science-and-technology revolution creates a revolution in the conversion of materials, energy stores, and – especially – in huge flows of information in the biosphere. Recently (exclusively, after the invention of electricity, electronics, computers, and planetary-information networks), this transition went to a qualitatively new state, the noosphere (from the Greek νοῦς – nous- "mind"), i.e., the sphere of brain activity and human thought (Vernadsky 1978). LIM acts as a geological force on the planet. This force acts more and more vigorously at all time–space scales around and inside our bodies.

The importance of a systemic (holistic) approach to LIM - a specific object of nature distinctive from any other kind of material object that one can meet on our planet - was recognized many years ago. Several traits (i.e., material growth and destruction, energy transformation, thermodynamics of such matter, and intrinsic connections between living and nonliving material components) are well known (Stephens 1888; Loeb 1906; Vernadsky 1929; Krebs & Kornberg 1957) but seldom formulated. Look, for example, in Google, how many papers you can find that are concerned with living matter? Strangely, modern authors seldom mention the fundamental object surrounding us on all sides. Several fundamental papers were written by Vernadsky (1929, 1978) but they, too, were rarely developed further, though recognized by specialists studying the biosphere. A very general and impressive description of LIM presented it as a living vortex. The hierarchy of metabolic recycling flows, resembling a huge vortex or whirlwind, was described some two centuries ago by the great French biologist G. Cuvier (1827). The most important properties needed "to form the just idea of the nature of life" are:

Life is a vortex ("tourbillon vitale"), the action of which is constant, all particles are in a state of perpetual mutation, constantly coming and going. The whirl action is constant, with all particles in a state of perpetual mutation, constantly coming and going. This vortex consists of combinations existing in one determined form, enduring for a limited interval of time. Existence of such combinations is supported by the continuous attraction of extraneous substances from outside, and by returning their own particles to the surroundings. There is a correlated ensemble of interdependent fluxes and huge mobs of relatively small particles. The system is alive only while this movement continues! This 'vital motion' prevents dissolution of the system due to the chemical affinities of the elements composing it. To penetrate the living body, all the elements are associated with the fluent liquid. The movement of liquids is possible only due to the existence of some arrangements or necessary conditions. A change in any of these conditions (partial movements) inevitably leads to cessation of life (Cuvier 1827).

This description was developed further by Vernadsky (1929, 1978), who suggested living matter - i.e., the whole ensemble of living organisms - as an efficient concept for studying the biosphere and ecosystem-level phenomena. He described living matter as specific gas-like matter: The rules describing living matter, being the rules governing the multitudes, may be analogous to the rules of gas, but must be essentially different from the rules governing a living organism that parallels a molecule of this gas (Vernadsky 1978). Recently, mankind is much more active and efficient, though not always more conscientious. The local ecosystems and the whole biosphere become more and more polluted and unsafe. More heightened awareness is needed to preserve the biosphere, i.e., our joint household (Vernadsky 1929; Khailov 1994).

Thanks to the global ecological crisis, we have begun to recognize that life is a planetary phenomenon. Life is the most powerful of all geological forces (Vernadsky 1978). Living matter exists and operates everywhere, from the top of the atmosphere to the bottom of the abyss, a 20-km–thick band that thrives around the Earth's surface. The understanding of the principles of the life-supporting processes is fundamental to our own survival. New approaches include varieties of sensors and data-processing methods to seek past and present life elsewhere. Recently, numerous institutes for advanced concepts have been developing more new approaches to biological studies (Margulis & Rummel 2001).

The minimal cell and minimal genome are already among the notable objects of scientific studies discussed and analyzed (Hutchison et al. 1999; Forster & Church 2006). Minimal models of the biosphere, ecosystems, and of living matter as a whole, are critically important for ecological studies. One such model was called the 'ideal minimal ecosystem' (IMES). A minimized set of keystone concepts describing living systems, ordered into a tree-like structure, can serve as a specific introduction to the thick forest of ecologic modeling.

THE CORNERSTONE CONCEPTS ON LIVING SYSTEMS (SUMMARY)

In order to produce a top-down decomposition of a complicated object (ecosystem) from a field of science (biology) troublesome for mathematical modeling, we developed a somewhat primitive (crude, but compact) condensed summation (a tree-like structure) interrelating the cornerstone concepts pertaining to living systems.

1. Of all the material objects known to science (gas, liquids, rocks, plasma, etc.), we can distinguish a special kind of material object, i.e., living systems.
2. Living matter (LIM) is a very special, discontinuous (divided into fragments), powder-like substance. The living matter consists of multitudes of living particles called organisms.
3. There is a strong distinction between two types of matter: living matter and inanimate matter. The quality that distinguishes living matter (LIM) from inanimate matter (IM) is called life.
4. Life is a complex process manifested in metabolism supporting several interrelating functions (growth, reproduction, response to stimuli, homeostasis, and adaptation to the environment, etc.). Hence, life is a specific temporal structure, a complex of metabolic transformations creating and supporting several types of matter, especially, living organisms and non-living organic matter (OM).
5. Word 'organic' means matter derived from living organisms (mainly, carbon compounds), relating to living particles or resembling them (synthetic chemicals that imitate natural OM); other types of matter are called inorganic.
6. LIM is corpuscular, i.e., it is a countable mob. Properties of a mob depend on its size. One living organism (individual) is a single being as distinct from a group. The rules describing living matter, being the rules governing the multitudes, may be analogous to the rules of gas, but must be essentially different from the rules governing a living organism that parallels a molecule of this gas.
7. An organism consists of one or more cells. Multicellular organisms are organisms that consist of more than one cell, in contrast to unicellular (single-celled) organisms.
8. A living cell is a special particle composed of several layers, including the closed membrane envelope and the inner part (protoplasm – a slimy, granular, " semi-fluid" substance) that contains water, the hereditary information (genetic materials, DNA and RNA) and several types of inclusions supporting metabolic processes. Some inclusions (organelles) also are protected by closed membranes. Hence, the living cell is a complicated system, highly different from any kind of lifeless matter. A lifeless cell resembles a living cell; the cell viability can be checked by a membrane permeability probe capable to distinguish between two types of matter (living and lifeless).
9. Each living particle has a limited longevity, i.e., the time range between its emergence and the end of process of life, called the death (i.e., a failure to perform at least one of key functions of life). After the death the former living particle loses some basic properties of life, stops to be living, converts into a dead organic particle. Dead particles and ruined organic structures pertain to inanimate organic matter (IOM). The criteria to distinguish between living and dead particles are sometimes

very complicated, especially for large multicellular organisms. Often an organism is a combination of living and dead cells; hence, it comprises the OM-particle with a high concentration of LIM. Some types of living particles (mollusks, corals, plants) are characterized with a low concentration of LIM.

10. Each individual organism emerges as a product of an act of reproduction. Reproduction is the key biological process by which new individual organisms – "offspring" – are produced from their "parents."
11. Cells are the smallest unit of life that can replicate themselves independently.
12. Reproduction is a fundamental feature of all known life. Some important organic structures (e.g., viruses) are incapable to independent reproduction. Several types of reproduction processes are known. The most interesting (i.e., sexual) type of reproduction process is via the formation of a zygote, i.e., the application of one reproductive cell transformed by a special procedure – fusion with a sex-partner cell - to generate a new combination of genetic material from a couple of reproductive cells. Genetic recombination produces offspring with combinations of traits, dissimilar to any parent. After that a zygote (one cell) begins a new ontogenetic cycle creating a living organism.
13. Potency of a cell specifies its potential to differentiate into numerous cell types. Totipotent or omnipotent stem cells can progresses into much larger systems composed of many different sub-structures (organs composed by different tissues of numerous cells including specialized sub-cellular structures). Omnipotent cells can develop into complete, viable, highly heterogeneous systems (multicellular organisms).
14. Thanks to recombination and spontaneous hereditary information changes (mutations) huge variability of organisms and specific cells was achieved. Multicellular organisms can contain a wide range of different cells, some of them being reproductive cells. A large organism can produce tiny reproductive cells that become zygotes of new large organisms, closing a new ontogenetic cycle - the key process of LIM temporal structure.
15. Many living particles (cells, colonies, organisms) do not produce any offspring. While all living particles have a limited durability, some organisms and cells succeed to produce offspring, and myriads of living cells continue the eternal process of life. Myriads of closed loops (ontogenetic cycles) create a very special kind of turbulent mass, i.e., a living whirl (LIW).

Chapter 4

IDEAL MINIMAL ECOSYSTEM

INTRODUCTION

The model, the Ideal Minimal Ecosystem (IMES), is a closed volume (buffer) protecting a closed recycling flow that implements the eternal (quasi-cyclic) process of renewal of flow-through elements (FELs) resembling the same (i.e., hierarchical) structure (Figure 1).

The main property of IMES is its ability to exist, i.e., to maintain its principal functional characteristics, withstanding changes in the environment, deficiencies of its composite materials, structure and subsystems, and faults in its inner components. This ability (or functional stability) is implemented through continuous replacement of numerous flow-through elements of the same (i.e., hierarchic) structure, a self-stabilizing process maintained within a closed envelope (membrane) protecting a volume of sheltered medium. This membrane, i.e., a closed thin layer of matter, provides insulation or isolation of all inner components from external impacts. Very often such membranes provide not complete isolation, but rather selective admittance of several resource flows needed to the inner machinery, obtained from the outside. The inner structures can also be active, capable to change the internal and outer space. The inner matter consists of liquids, gels, and numerous disperse components. Dispersed constituents consist of multitudes of very small (in comparison to the ecosystem size) FELs. Not all FELs are permanent; a large part of them are dynamic; they go through an ontogeny cycle: fission, growth, development, search for a vacant place in the system, adaptation to the local conditions, changing the way and efficiency of functioning; offspring production; aging of inner substrates, structures and algorithms of functioning; growing incapability of self-maintenance or stabilization, incapability to counteract changes in their inner substrates or surrounding environment; the final transformation is the FEL destruction.

As each function of IMES is performed by multitudes of FELs, inclusion or exclusion of some components, even of a large part of them, is not crucial for the system as a whole. Such problems are solved via obtaining resources and production of new FELs, i.e., successors to the previous FELs. A very important property of the whole system is the almost complete resource regeneration that solves both the problem of resource deficiencies for production of new FELs and the problem of the "burial" of the eliminated FELs (Kamenir 1983).

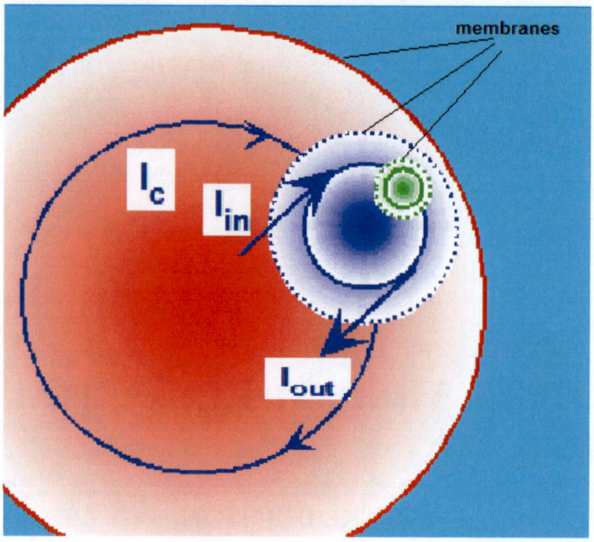

Figure 4-1. Ideal Minimal Ecosystem (IMES). Iin, Iout, Ic - in-, out-, recycling flow, respective. Note the hierarchy of membranes producing the hierarchy of spaces and recycling flows (after Kamenir 1986).

Ideal Object

As IMES is an ideal model, it depicts not all real complications but rather a minimal set of interrelated problems and solutions of principal importance.

The most important features of IMES (Figure 1) are:

1. the implementation of the whole system via ephemeral flow-through elements;
2. the resource set recycling as a means of resource supply and pollution prevention;
3. distributed structures, i.e., a way of realization of each function by a high number (i.e., a mob) of FELs;
4. a specific way of realization of each function by a heterogeneous ensemble of FELs going through different stages of ontogeny cycles (again, distributed structures, heterogeneous mobs);
5. the FEL inner-medium stabilization with the help of membranes, i.e., active control of the border-layers and border-crossing fluxes;
6. the feed-back control provided by some of closed loops;
7. the application of resonance, i.e., a highly effective mechanism of selective stabilization of time-space scales, intrinsic to closed membranes and closed loops;
8. the use of hierarchic structures within FELs (particles, membranes, processes, and networks);
9. a broad spectrum of time-delays through the hierarchic networks of the closed loops, as an effective means of feed-back stabilization (Kamenir 1986, 1987).

The IMES model is minimal in the number of elements and features applied to describe the principal properties. Many its properties are produced as by-products of principal elements and traits selected; e.g., the feed-back can be produced by some loops, what is typical for recycling, i.e., for recycle nets; the resonance is a property of closed spherical envelopes; the distributed delays can be produced by heterogeneous mobs that are formed by environmental noise and random impacts causing the lack in precision of the new FEL-copy production. While the system (IMES) is ideal, it does not describe a real object but rather simplified structures and asymptotic states, such as a geometric point (i.e., a zero-volume object), a zero-width line, the absolute zero temperature (-273°C = 0°K), the ideal gas, the perpetuum mobile, etc. (Nicolis & Prigogine 1989).

A real system may move towards such an ideal state, but can never achieve it, due to the existence of additional factors not considered in a minimal model. However, the existence of a known asymptotic value (a specific point, line) can be used for the development of systems of coordinates, measurement of certain distances, comparisons of real objects and quantitative analyses of differences and displacements within the coordinate space. Hence, such an ideal estimate can be used for quantitative description of the system state, progress and evolution.

Selecting the self-maintaining recycling movement as the main property of the object, we obtain a hierarchy of closed recycling fluxes and flow-through elements (Figures 1 and 2).

There are a number of models, described in recent times by other authors, that considered several peculiarities applied in IMES: continuous flow-through movement (Cuvier 1827), recycling flows (Finn 1980; May 1981; Patten 1985), feed-back loops (Stone & Berman 1993), hierarchy of cyclic blocks and processes (Lotka 1925), and dynamic ensembles composed of huge mobs of molecules of a gas-like living matter (Vernadsky 1929, 1978).

In this way, IMES depicts not only a hierarchical structure of natural ecological systems but also the biosphere as a whole, the highest level ecosystem of the earth (Odum 1971), and the intracellular cyclic nets of the metabolic reactions. As such, it tries to integrate concepts of the biogeochemical cycling, the living matter of V. Vernadsky (1929, 1978) and the Tourbillon Vital (a living whirl) of G. Cuvier (1827). The Earth biosphere looks as the most natural example of the ideal ecosystem, i.e., a highly isolated sphere of life, separated by tremendous distances from other such systems, and producing very small exchange flows with other inhabited objects. Modern astronomy provides evidence to the existence of numerous exoplanets (i.e., planets orbiting stars other than the Sun) with Earth-like traits. A recently emerging point of view is that numerous planets, even those very different from Earth, may have the right conditions for life (Seager 2013). One can suppose that some of such planets can be inhabited, i.e., support life known to us or in a somewhat different form. Therefore, it is reasonable to consider not the one-sided hierarchy (i.e., top-down, going only down from the largest ecosystem, the biosphere), but also include the opposite direction, going upward from our planet to the solar and interstellar scales. Typical islands of life should be galaxies including myriads of planets, and connected with other galaxies and metagalaxies through tremendous spaces by sparse seed flows of life, e.g., exo-microorganisms in meteorites and space ships (Yang et al. 2009). In our biosphere, living matter covers almost all surface of our planet by very thin layer condensing energy store (mostly accepted from the Sun). The LIM energy involves a store of chemical elements in unceasing cycling movement supporting the existence of all particles of this eternal structure transforming and protecting the planet (Vernadsky 1978).

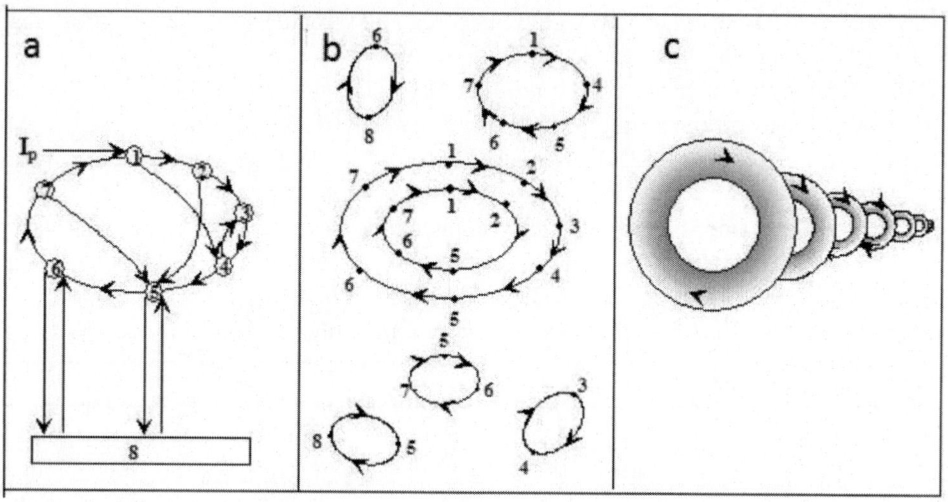

Figure 4-2. Transformation of the trophic net (a) to a cascade of correlated recycling fluxes (b, c). Ip - primary production, i.e., energy input ((after Kamenir & Shteinman1998).

IMES resembles also a broad class of objects studied in recent times by a rapidly developing branch of science, i.e., the thermodynamics of irreversible processes or the theory of dissipative structures. The optimal approaches to analyze structure and dynamics of such objects are statistical ones (Nicolis & Prigogine 1989). Such structures are, as a rule, very different from living organisms, but some traits of such abiotic objects (e.g., tornado, water fountain, sea waves, etc.) strongly resemble living systems (they emerge with the flow of energy, they 'die' if the energy is finished; they maintain the same shape while the involved particles or drops are changed, drops come in and go out continuously, etc.). All such systems (including living and abiotic ones) resemble cascades of water and whirlpools described above by Cuvier (1827).

Chapter 5

MODELING AND MINIMAL BUILDING BLOCKS

With the help of the 'Living Vortex' concept, numerous inhabited territories of our planet and the whole our biosphere can be described as parts of an eternally spinning whirl of living matter entailing numerous living particles and various non-living components. The LIM is just one of several types of matter, but it is the most important and extraordinarily active. Such kind of matter has its specific traits, needs, preferable conditions, and maintenance problems. Living components can be studied with the help of general sciences (physics, chemistry, mathematics, etc.) but often specific –'biological' - methods and schemes help us understand, monitor, modify, and restore living systems much easier and efficiently. Look at agriculture, forestry, at a large orchard, or at a small kinder garden. For example, you should frequently add water, fresh air, biogenic chemical elements (fertilizers, nourishments or specific food), take out excessive heat outflow, CO_2, and other consumed constituents. Many kinds of LIM produce flows of substance and information (litter, excrements, songs and noise, flowers, fruits, etc.); all LIM produce heat, i.e., energy dissipation flows.

The temperature dependence of living organisms is of key importance. Its connection with physical laws of nature is understood, and there are numerous biological mechanisms modifying processes important for the survival of living systems (Begon et al. 1996; Koga 2012). Important physical characteristics of each system, including any living system, are spatial and temporal dimensions. For a mortal particle of LIM, they are the longevity, size, volume, and mass. A principal characteristic of each living system is its weight (mass and respective volume of material entities). Relation between the LIM weight and quantity of particles comprising the system depends on the particle mass or size composition of such material bodies. As a rule, there are particles of different size; therefore, connections between the living system energy dissipation (respiration, R), its biomass (B), and quantity of comprising it particles (abundance, N) are not very simple. Recently, two general tools become popular, i.e., the analysis of particle size distribution (the biomass size spectrum, BSS) and regression equations connecting various traits of particles.

The proportionality between the biomass and its metabolism strongly depends on the particle size composition. The logics of such proportionality can be easier grasped with the help of geometric studies demonstrating complicated internal spatial structure of a living body and its external surface. Flow of various components through the body and its external surface (S) depends on the body volume (V) and its surface-to-volume ratio (SV). Connections between various traits of the living body are, as a rule, non-linear, but using logarithmic data-transformations very simple regression equations – allometries – can be found (Peters 1983). Now, allometry is a vast field of science that analyzes the living body morphology, including the body shape effect (Mandelbrot 1982). This morphology (e.g., an egg, a serpent, a tree, a coral, a sponge) explains often how the living body volume, surface and metabolism scale with the body size. The number of studied traits and respective allometry equations is very impressive (Peters 1983).

Connection between various traits of living systems, their static and dynamics comprise a huge field of science – modeling (physiological, ecological, biological). People speak about a special activity field, i.e., the art of modeling producing various types of models (Odum 1971). Special tools and methods were developed supporting modeling of living systems, e.g., the general theory of systems (Bertallanffy 1952), various mathematical methods and quantitative generalizations independent of the discrete mechanisms in living objects (Murray 1926). Several optimization principles were formulated, including the Ockham's razor principle (Lat. lex parsimoniae, i.e., the law of economy of thought). This principle states that 'entities and relations must not be involved beyond necessity.' In other word, the number of parts, links, and traits described should go to the minimum. Such models are called minimal, and, going to modeling, we can see the word minimal ever more frequently, for example, the minimal cell (Forster 2006), the minimal genome (Hutchison et al. 1999), and the minimal ecosystem (Kamenir 1986).

A system (and its model) can be minimized or non-minimized, i.e., to have excessive (doubled, multiplied, redundant elements). Trying to minimize the studied biological systems, modern techniques were developed allowing us to find even excessive molecular components of genome, i.e., components unnecessary for survival in the chosen laboratory conditions. An important question occurs: how many genes are essential for cellular life of the living system and what is the fraction of non-minimal components (Hutchison et al. 1999).

Inclusion of additional elements can provide excessive guaranties of functioning; but to understand clearly the principle of functioning, the system should be purified from all redundant or superfluous elements. Then we can find a more general, less complicated or more efficient solution. For example, we can say that a 'clock' is a tool transforming an interval of time into digits (a counter of units of time). So, an electronic clock needs just a generator of oscillations and a counter. But to make this count visible we need also an indicator or a clock-face. Really, the counting process demands some energy, therefore, an accumulator is necessary for a material tool applicable for reasonable time intervals (days, years). Another type of clocks can be a sandglass (a sand timer), that measures the passage of sand grains. Generalizing, we can say that a clock needs a quota of material elements (a stock of particles or fluids – water, electricity, or grains), plus a couple of canisters, and an hourglass, to produce the estimate (Figure 1); a set of such components can produce a materializing (i.e., emergent) quality, the visual estimate.

The principal element there is a specially constructed hourglass indicator built on the basis of two glass cones (Figure 5.1). The unit time interval (e.g., one minute) transports the same quantity of matter (sand), but the vertical distances on the indicator, corresponding to the same time interval, are variable, they depend on vertical positions of each mark. We can transform a sand-timer to a clepsydra (a water-timer). Then we replace the sand by water, or by wine, oil, benzene, even by melted metal. The emergent property (the time interval measurement) is maintained, but only with the help of a special formula for the indication steps. The time estimation was read from graduated lines on the cone. Further, the Romans invented a better indication system, a clepsydra with a cylindrical vessel where a float demonstrated readings against a scale on the cylinder wall. Cylindrical tanks produce a linear scale with uniform vertical intervals for the same time interval (compare Figure 5.1B and C). Then the time scale creation becomes simpler. We can also add many interesting elements (e.g., a box, an operational manual, price indicator, spare canisters and spare liquids, etc. All additional components can make a system more efficient, less expensive or more precious, but they do not change the system (the clock) principal function.

The same situation can be seen in IMES or LIV models: the FEL with ontogenetic cycle is the basic invention capable to replace an ephemeral part of each living system by its principally-similar progeny-part. After it, a cycling offspring production provides a very long (as we see, practically eternal) functioning of the vortex. The LIM exists thanks to replacement of FELs – cells, colonies, organisms, and other living systems. In general, LIM supports life (a process of living system functioning) via a special time-structure (cycling), providing replacements of LIM particles.

All living systems are made of cells. There is at least one living cell in a living organism, but very often there is much more than one cell in a FEL (colony, organism, and ecosystem). A multicellular FEL can carry on a number of functions, and just opposite situation is common: many living systems are built as cooperation of highly specialized mobs or huge replacement stocks of almost similar elements (cell cultures, colonies, tissue, populations). The cell is a basic building block, capable to maintain a quota of genetic information and use it, to control metabolic activity, i.e., to accept material-energy-information stocks from its surrounding, transform such stuff and excrete some by-products back to the surrounding medium. All cells come from the division of preexisting cells using genetic material (DNA, RNA) that is passed down to daughter cells from its mother-cell. Therefore, each FEL (a living particle) has two time borders (the beginning and the end).

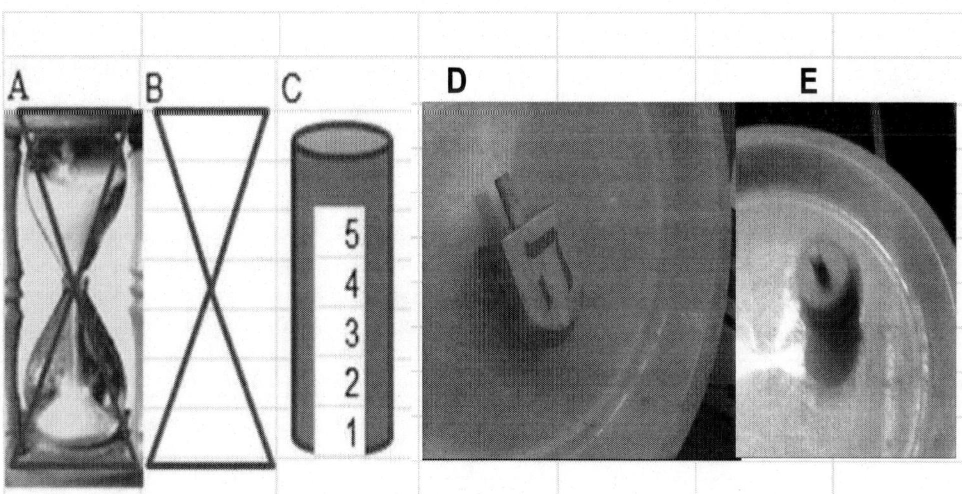

Figure 5-1. A sand-clock. A, B is the glass cones and the variable-scale formal scheme; C – cylindrical container with a linear indicator-scale; D, E is a small modification changing the key traits of a system (i.e., a spinning whirl): LIM is not any mass, but *a moving mass*.

Genetic information stored by each cell supports also the functioning of the cell itself. So we can select a genome (i.e., the complete set of genetic material needed to create and maintain an organism) and protecting it membrane envelope as a minimal building block of LIM, drastically different from any other kind of material objects known. We can find several more very general components (e.g., mitochondria support the energy supply, photosynthetic pigments support the most important energy influx provided by our Sun, ribosomes transform the genome information into material complexes of proteins, vacuoles contain spare materials, etc.); all such parts upgrade the cell functioning not changing the emergent essence of life, interpreted here as the eternal movement. Such movement is produced by ontogenetic cycles (generations) of flow-through cells. Hence, the first two components (the genome and its membrane envelope) can be selected as the minimal complex, the basis of each living system. We can imagine a living system as a 4D-object with a cyclic sequence of several ontogenetic stages.

An intrinsic condition of the LIM existence is the regular update of genetic material carried on with new copies of cells; the ageing mother-cells doomed to be ruined in order to give resources and liberate space for their successors. There is an eternal recycling movement needed for and based on the eternal update of almost eternal store of genetic information. The ideal system would have the perfect (100%-true) reproduction of the genetic information; in real systems the DNA-replication is just a molecular copying process and sometimes there can be errors. Such errors (mutations) can lead to the death, or survival of a modified genome. Such process produces daughter-cells that are a little different; thanks to such process (the flow of mutations), we see evolutionary dynamics and various organisms are produced by the same predecessors. While such biological mechanisms are very efficient - almost ideal - one error to billions of replicating acts (Keightley et al. 2015) - we need mathematics to understand the consequences of astronomical numbers of elementary acts. If the replication time is several hours (for many microorganisms) and there are several million of such organisms in each liter of water or soil, and there are billions of billions of liters of water, and such mutations continue during several million (and even, billions) years, such analyses are very complicated. Special kinds of biological calculations (mathematical models) are necessary.

Chapter 6

ALLOMETRY AND METABOLIC SURFACE

INTRODUCTION

Exchange of gaseous and dissolved substances with the environment is one of basic conditions of life. Such exchange is performed via several types of surfaces (cell surface of microscopic algae, leaves of higher terrestrial plants, skin, gills, lungs of animals, etc.). One can note that the quantity of energy needed daily to a living organism is highly variable and depends on several traits of living organisms. Thanks to the tremendous range of body sizes it is easy note that the quantity of food and energy respired depend on the organism size or body mass (Peters 1983). For various animals and plants, the energy exchange - respiration and photosynthesis - can be measured via gases - oxygen (O_2) and carbon dioxide (CO_2).

The ratio of respiration (R) to body mass (W) was studied for numerous species of animals across large body mass ranges (Hemmingsen 1960; Muir 1969; Peters 1983). Much smaller attention was given to plants (Khailov & Parchevsky 1983; Karmanova & Ilina 1984; Moloney & Field 1989; Joint 1991). Therefore, we made an effort to calculate the relation between the plant external surface (Sg) and its body volume (V) in a maximally broad range of plant sizes. Such a regression equation could be compared with respective allometries for metabolism and body surface of various kinds of plants, animals and other organisms.

MATERIALS AND METHODS

The main part of data was obtained from field studies on marine macrophytes at littoral of Black, Barents, Japan and Okhotsk Seas (shores of Russia and Ukraine) by several marine biologists under general supervision of K.M. Khailov (Firsov 1979; Khailov & Parchevsky 1983; Khailov & Firsov 1985). The dry and fresh wet were recalculated to the body volume, using coefficients of 5:1 for fresh-to dry weight and 10^{12} um^3 for 1 g fresh weight (Khailov & Kamenir 1988). Several extremely small species - cyanobacteria and viruses (Mendzhul et al. 1985) - were added to analyze the small-side region of the regression, expanded further with unicellular aquatic algae estimates of Raymont (1980). Higher terrestrial plant estimates from review of Karmanova & Ilina (1984) help us to continue the size range to the right, up to the extremely large trees.

RESULTS AND DISCUSSION

Unicellular plants (30 species) encompass some 11 orders of magnitude of body mass. Then, green, brown and red multicellular algae (15 species, 495 measurements) covered another 6 order of magnitude of the plant fresh weight. Higher terrestrial plants (26 species) added some 12 order of body mass, partially overriding the aquatic macrophyte size range. The total body mass range for the regression comprised about 24 orders of magnitude.

Parameters of allometric equations were estimated with the help of general linear regression base on logarithmically transformed values:

$$\log S,\ \mu m^2 = a + b \log V,\ \mu m^3, \text{ where } a \text{ and } b \text{ are coefficients.}$$

For several regression equations, considering various groups of plants, the estimates were highly variable (depending on the organism species, habitat, season and other environmental conditions). For trees and whole macrophytes, coefficient b = 0.534 to 0.927; extreme coefficients (b>1) were obtained for small branches of macrophytes. The general regression (24 orders of the body mass range) produced $a = 0.652 \pm 0.344$ and $b = 0.779 \pm 0.008$ (Khailov & Kamenir 1988). Therefore, the slope coefficient of the general regression was considerably different from traditional geometrical models (log S ~ 2/3 logV) (Rubner 1883; White & Seymour 2003). Very interesting was that this general SV-slope was rather close to the highly popular estimates of 3/4 (i.e., b ~ 0.75) cited for respiration and several other traits of living organisms (Peters 1983). The correlation between the main metabolic processes of autotrophic organisms (photosynthesis) and the active surface of plants resembles the correlation between energy dissipation and respiration surfaces of animals. The regression slope for surface of plants (0.779) is very close to slope estimates for metabolic expenditures, discussed in numerous publications (Peters 1983; Schmidt-Nielsen 1984; West et al. 2005).

The energy fluxes must be carried through all living parts of each living body; this flow should reach every tiny element of living flesh (i.e., living cell). The necessary amount of energy should enter the living particle through its external surface, going through branching capillaries up to small structures of the living substance (living cells and its organelles), and then, the energy dissipation flow of the same level should go back, crossing the external surface of the living particle and go out to the living particle environment. Hence, if the SV-ratio of a living body diminishes due to the body volume growth, the same trend (diminishing RV-ratio) and similar regression slopes can be expect for the growth processes and for respective metabolic energy fluxes. Hence, we can expect close regression slopes for the body metabolism and for external surface estimates of living organisms. This correlation should be produced via a special structure of the body surface formed by goffers and several levels of branching, i.e., via a special -fractal (Mandelbrote 1982) – organization of the body surface. A general conceptual approach (i.e., transport nets of metabolic substances) was developed (West & Brown 2005) providing reasonable explanation for general traits of LIM particles of varying size.

Oxygen transport and expenditures of living organisms are among the most advanced fields of quantitative physiology and theoretical biology (Murray 1926). Some quantitative generalizations have been proposed that are independent of specific mechanisms. The principle of the maintenance of steady states furnishes definite equations indicating equality of intake and output by elementary living structures. Another general and very helpful principle relevant to physiological systems is the principle of minimum work, concerning the operation of vascular systems transporting blood (i.e., carrying oxygen that means the energy flows) in living organisms (Murray 1926). Despite amazing diversity and complexity, many fundamental biological processes manifest an extraordinary simplicity if described as a function of living body size.

A set of principles was proposed based on the observation that almost all living systems are supported by hierarchical branching networks with invariant terminal units, space-filling and optimized for the cost minimization of the transport work. These general constraints lead to a quantitative, predictive theory portraying many essential features of living systems. Such models considered animal circulatory and plant vascular systems (West & Brown 2005). The model should be viewed as an idealized representation ignoring numerous complications, e.g., turbulence, nonlinear effects, etc.; such second-level details play only a minor role in determining the model principal properties. Biological networks vary in the properties of the tubes, the fluid transported (liquid to gas), and the nature of the pumping mechanism (a pulsatile compression pump, diffusion in insect tracheae, osmotic, vapor pressure). Despite numerous differences, such space-filling networks reveal essentially the same scaling laws. A space-filling fractal is a natural structure ensuring that all cells are serviced by capillaries. In order to minimize the energy dissipated in the system, the network must be a conventional self-similar fractal. Just as the IMES, the model of West and co-authors considers the transport nets at several level of hierarchy up to mitochondria and respiration molecular complexes (West et al. 2002; West & Brown 2005).

Hence, the LIM is a metabolizing mass supported by a transport net; this net carries materials and energy, and takes back the decayed parts to be thrown out. Such work involves metabolic expenses (energy dissipation) that make the LIM dependable on the external energy. Therefore, the basic drawback of IMES (the energy dependence) is to minimize the energy dissipation.

There was also an inverse relationship found between the size of an animal species and its local abundance (Damuth 1981). Density of living communities (organisms per square kilometer) is related approximately reciprocally to individual metabolic requirements, indicating that the energy used by the local population of a species in the community is independent of its body size. It can be a general rule of community structure. The widespread occurrence of an approximately reciprocal relationship between population density and individual metabolic requirements among mammalian herbivores suggests that a general principle is involved. The independence of species energy consumption and body size revealed by this reciprocal relationship implies that random environmental fluctuations and interspecific competition act over evolutionary time to keep energy control of all species within similar bounds (Damuth 1981). Hence, we can expect similar bounds for energy consumption values by LIM and similar bounds of the LIM particle external surface (m^2 per m^2 of the planet surface) (Kamenir 1992). If the energy input (the sun light) is roughly the same (W per m^2 of the Earth surface) at vast areas of the Globe (excluding its circumpolar zones), we can expect roughly same values of integral estimates of LIM, such as the LIM respiration (W per m^2) and LIM integral surface index (m^2 m^{-2}).

Chapter 7

SIZE SPECTRA AND TOTAL SURFACE AREA OF LIVING MATTER OF THE OCEAN

INTRODUCTION

A basic condition of existence of each organism is the metabolite exchange across its body surfaces and its environment (Dol'nik 1978). Vernadsky (1978) regarded the individual and aggregate surface area of all living community as one of the most important characteristics of our biosphere. Such ideas of Vernadsky were appreciated after a very long delay. Recently, ever more popular becomes his concept of *biosphere* as a thin 'film' of living organisms covering the Earth. As a rule, the data describing living organisms of the Earth are formulated in terms of biomass (B) or abundance (N). Given the huge range of organism body sizes (approximately 10^{-8} to 10^{2} m), the assessment of various size groups of living organisms changes if the community biomass, respiration, abundance of organisms, and surface area of the living particles are compared. Application of biomass estimates results in exaggeration of the role of large organisms, whereas the use of abundance figures exaggerates the importance of small organisms (Odum 1971).

Applying one of the simplest geometric approximations (i.e., assuming the body shape to be spherical) we obtain

$$W_t = \rho \pi D^3/6 \qquad (1)$$

$$N = B/W_t = 6 B D^{-3}/(\rho \pi) \qquad (2)$$

where W_t is the body mass, D is its effective spherical diameter (ESD) determined from the body mass, and ρ is the organism specific gravity ~ 1 g cm^{-3}. A possible intermediate case between B D^0 and B D^{-3} is the total surface area of a group of organisms

$$S_{ts} = N S_t = 6 B D^{-1}, \qquad (3)$$

where

$$S_t = \pi D^2 \qquad (4)$$

is the individual body surface area. The allometric relationships between various metabolic characteristics and the organism body mass, obtained for numerous organisms, look as: log(R) ~ b log(W_t). The exponent b is generally from 0.6 to 0.83 (Dol'nik 1978, Peters 1983), indicating a closer correlation with S_t (b ~ 0.67) than with W_t (b ~ 1).

Current bio-oceanographic data support the ideas of Vernadskiy, considering the importance of the dispersed state of living matter and its surface area; such correlations can be illustrated quantitatively for many groups of organisms of the world ocean. To begin, let us compare the most general functional characteristics of the main groups of the ocean organisms, i.e., production and respiration with other traditionally cited quantitative measures - N, B, and S.

MATERIALS AND METHODS

The idea of such comparison (Kamenir & Khailov 1987) was based on data of Sorokin (1978) who has compiled a set of total annual figures for the World Ocean taken as a whole. Such set enabled us to eliminate the effects of seasonal and geographical variability, spatial and temporal variability, vertical stratification, and migration of organisms. The data for the individual groups of organisms were assigned to the organism size scale by means of the classification table of Sieburth et al. (1978), pico, nano, microplankton. For benthos, which was not included in this table, we applied the size range 0.1-10 cm. The plant group was divided in two: phytoplankton (1—100 μm) and phytobenthos (comprised by microphytobenthos and macrophytes) ~ 1 μg to 1 kg of live weight (ESD = 100 μm to 10 cm). The biomass was divided equally between phytoplankton and phytobenthos, while the production was divided in 20:1 ratio (Whittaker & Likens 1975; Shushkina et al. 1984). In this way we constructed (Figure 1) size spectra (SS) producing the percentage distribution of the total value of several characteristics (total biomass, total body surface area, production, and community respiration) over size classes.

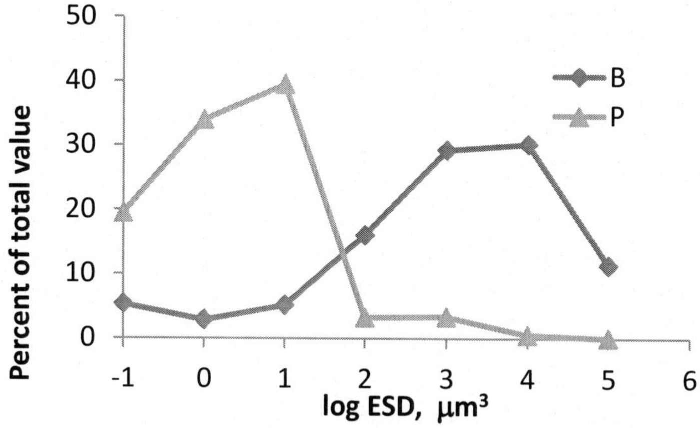

Figure 7-1. Size spectra of living matter of the World Ocean. Biomass (B) and production (P) of all organism (percent of total). Groups of organisms (Table 1) distributed according to the ESD size scale; based on raw data of [Sorokin 1978, p. 590].

Table 1. Groups of organisms according to log ESD scale (μm) and respective fresh weight

logESD, μm	0	1	2	3	4	5	6
Fresh Weight	pg	ng	mcg	mg	g	kg	ton
Groups of Organisms	Bacteria	Phyto-plankton	Phyto-plankton	Mesozoo-plankton	Mesozoo-plankton	Macrozoo-plankton	
			Microzoo-plankton				
				Microzoo-benthos	Benthos	Benthos	
					Macrophytes	Nekton	Nekton

Each size class represents an increase of an order of magnitude ($ESD_{i+1} = 10\ ESD_i$) in the organism geometric size (Sheldon et al. 1972; Peters 1983).

Sorokin did not present data on the virioplankton, and we also did not show viruses as 'living organisms' (see above). Thus, the left end of all size spectra was truncated. The breakdown of data produced seven size classes, each representing an increase of an order of magnitude in the organism ESD. When one group of organisms described by Sorokin represented several size classes, the respective quantity was subdivided between size classes in equal fractions. After that, for each (i-th) size class, the 'contributions' of the various groups (e.g., macroplankton, benthos and nekton in the 1-10 cm class) were summed and normalized to 100%, i.e., $X_{i\%} = X_i\ 100\%/X_\Sigma$, where X_Σ is the total value of the respective characteristic (e.g., biomass) for all living matter of the World Ocean. Starting from the biomass, assuming 1 kcal = 1 g live weight = 1 cm^3, we used equations 2 and 3 to calculate the abundance and the total surface area of living organisms.

Naturally, the real, dissected shapes of the living bodies not always are spherical; therefore, their surface areas - producing metabolic exchange with the surrounding medium - the living body outer surface (i.e., branches, leaves and roots, in the case of plants), and primarily internal exchange surfaces (i.e., lungs, gills, and the gastrointestinal trace in animals) are considerably greater than those determined by equation 4. Therefore, the notion of "metabolic surface area" ($S_{tm} = S_{lungs} + S_{GIT} + S_{skin}$) was introduced (where GIT is the animal gastrointestinal trace). The "total metabolic surface area" of a biota or any of its components was calculated: $S_{tms} = \Sigma\ S_{tm,\ i}\ N_i = \Sigma\ KSS_i\ S_{t,\ i}\ N_{i,}$ where *i*= -1 to 5. The 'surface area ratio' or the surface-to-surface coefficient KSS was determined in allometric form. Starting from bacteria (with KSS = 1, ESD = 1 um) we obtain for the Man (i.e., *Homo sapiens*), S_{tm} = 300 and 1.5 m^2 of the internal and external surface, respective, with Wt = 70 kg (Begon et al. 1996). Therefore, we obtain the ratio between the geometric and metabolic surface, approximated as KSS ~ $W_t^{0.14}$. Thus, KSS ~ aW_t^b, where *b* ~ 0.667 + 0.14 ~ 0.807, in a good agreement with the estimates of *b* exponent in other equivalent equations: 0.82 for the gill area of marine fish (Muir 1969) and 0.78 obtained by us (Khailov & Kamenir 1987) obtained from a broad size range of aquatic plants, supplemented also by published data (viruses, bacteria, unicellular algae, and leaf surface of terrestrial higher plants).

Results

Size spectra (SS) for several characteristics (biomass, abundance, production, destruction, total and aggregate surface area) of living matter of the World Ocean (Figure 1) were shown in Table 2 (absolute estimates, and percent distribution).

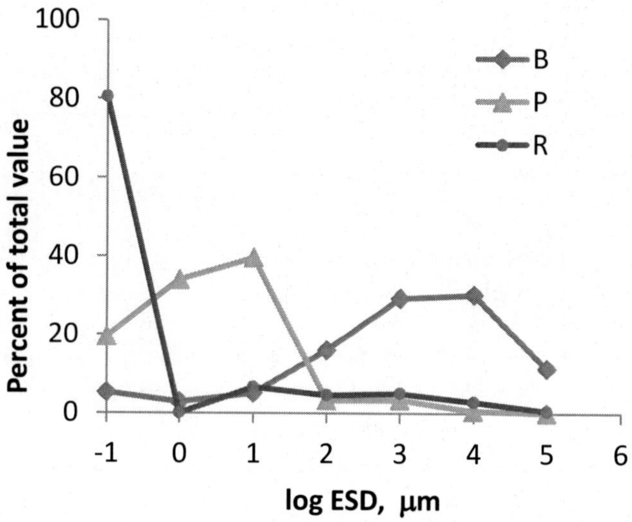

Figure 7-2. Size spectra of living matter of the World Ocean. Size distribution of total biomass (B) in comparison with metabolic parameters – production (P) and respiration (R) of all groups of living organisms.

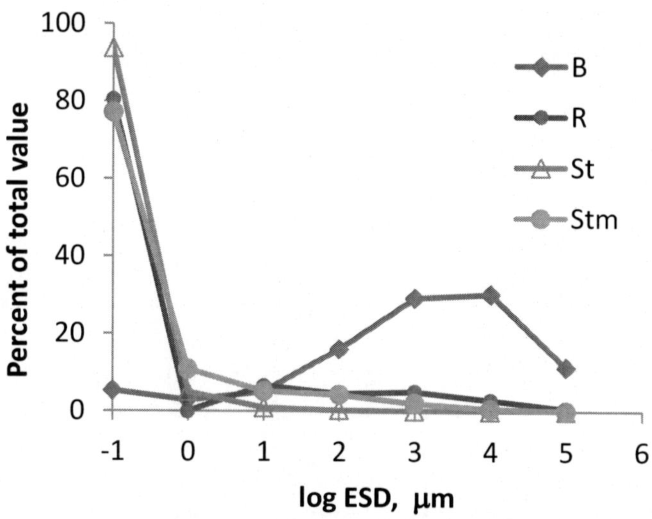

Figure 7-3. Size spectra of living matter of the World Ocean. Comparison of size distribution of respiration (R) and biomass (B) and external surface of organisms (simple geometric model, St, and fractal geometry corrected, Stm, i.e., allometric coefficient b = 2/3 and b ~ 0.79, respectively).

Table 2. Size distribution of living matter of World Ocean

	Size Fractions, ESD, um							
Parameter	**-1**	**0**	**1**	**2**	**3**	**4**	**5**	**Total**
B	1.4	0.75	1.35	4.25	7.75	8	3	26.5
log10N	28.93	25.66	22.91	20.41	17.67	14.68	11.26	28.95
P	210	366	426	35	36.05	5.65	0.6	1079.3
R	490		40	28	30.45	17.1	3.35	608.9
St	26.56	1.42	0.256	0.081	0.015	0.002		28.334
Stm	26.56	3.73	1.77	1.48	0.7	0.19	0.02	34.45
KSS	1	2.63	6.92	18.2	47.86	125.89	331.13	
P1	439.4	219.2	73.4	14	14			760
Pk	649.4	219.2	133.4	35	36.05	5.65	0.9	1079.6

Note: Size distributions (size spectra): biomass (B, 10^{15} kcal); logarithm of organism abundance (N); production (P) and respiration (R), 10^{15} kcal/yr; total surface area of organisms (S_{ts}) and total metabolic surface area (S_{tms}), 10^{15} m^2; S_{tms} apply surface-to-surface ratio (KSS). P1 and Pk are two variants of reasonable size distributions of primary production, 10^{15} kcal/yr (Kamenir & Khailov 1987). The size class number is the logarithm of the class effective spherical diameter (ESDm~$W^{1/3}$).

The size spectra for production (P), destruction (R) and aggregated surface area (S_{tms}) were qualitatively similar (Figure 1); both of them differ considerably from the biomass size spectrum (B). The man part (~90%) of the cumulative quantities of several variables (P, R, S_{tms}) was contributed by organisms smaller than 100 um, primarily phytoplankton and bacterioplankton. One should note that the data on these groups of organisms had a major reevaluation (Shushkina et al. 1984). Several important results considering microbial communities (Takahashi & Bienfang 1983; Laws et al. 1984; Zavarzin 1984) helped us correct the distributions of some parameters (Figure 1) in this part of size spectra. The prime production (P1) is divided between the 0.1, 1 and 10 um classes according with the approximate proportion of 60:30:10 rather than 0:50:50 applied when calculating plots of Figure 1A.

DISCUSSION

The main result of this study was a rather good agreement between the metabolically active surface area of the biota and its main functional characteristics (i.e., respiration and production); the agreement was quite satisfactory for the World Ocean as a whole. One should stress again that we speak about the metabolic surface area, producing metabolic flows between the organism and its environs, which is much larger than the living organism external surface. This agreement is particularly valuable because other methods of comparing these characteristics show considerable divergence. As examples of such divergence, we note the poor correlation between the functional characteristics (P and R) and the main traditionally applied structural characteristics (B and N) of various groups of organisms.

This agreement was achieved through a specific scheme (size spectra) representing the totality of organisms. Size spectra have already been used in numerous theoretical and experimental studies in ecology and hydrobiology (Sheldon et al. 1972; Barenblatt & Monin 1983;Takahashi & Bienfang 1983), including analyses of production and destruction processes (Tseytlin 1986). The body mass and body size are formal, quantitative features extremely important in biological studies, closely associated with the biological characteristics of organisms (Peters 1983). The body mass and its geometric size are efficient in mathematical transformation and mathematical modeling.

The novelty of this study was the comparison of size spectrum series for the totality of organisms of an ecosystem, i.e., its living matter taken as a whole (Figures 1 to 3). This comparison revealed that metabolism estimates (P and R) correlated better with the surface area (S_{tms}) than with the traditionally applied characteristics (B and N). The close correlation of production with S_{leaf} (S_{tms} for plant ensembles of various ranks) has been long ago known in ecology of the higher plants; it has been investigated (Karmanova & Il'ina 1984) and used for forecasting crop yields. The correlation between the surface area and metabolism has been discussed at various levels of organization of living matter, including the ocean as a whole (Ayzatullin et al. 1984). Of particular interest is the structural organization of the metabolic surface area, i.e., its fractal geometry (Mandelbrot 1982) or repetition of similar spatial structures at various size scales and levels of organization. Such spatial structures have been studied in considerable detail for marine multicellular algae (Khailov & Parchevskiy 1983). The general tendency of biological systems to place the maximal metabolically active surface in the given volume, that extends down to the molecular level (Barenblatt & Monin 1983), witnessed in structures typical of living material such as treelike dendrites (i.e., hierarchical branching) and nested membrane structures (i.e., hierarchical inclusion). The fractal structure of living matter increases its surface-to-biomass ratio and the aggregate surface area of oceanic LIM. In the context of Vernadsky's ideas it is important that the individual mass and specific body surface (SV) are indicators of the degree of dispersion of living matter, whereas the shape of metabolic surface (S_{tms}) size spectrum probably expresses types of fractal organization of metabolic surface. Such geometrical organization obscures comparisons of the surface area and its metabolism, as the evaluation of S_{tms} depends on the accuracy applied for approximations of the hierarchically dissected surface. At the same time, such hierarchical organization makes easier the surface estimation through application of identical formulae for ensembles of structural components. The principles of self-similarity are manifested in geometric plots approximating various sections of the biomass SS plot (Barenblatt & Monin 1983), and the S_{tms} spectrum as well. If self-similar structures are applicable when comparing size spectra describing objects situated at various hierarchical levels, then mathematical methods can be used to reconstruct and evaluate some regions of size spectra.

It is possible that the principles of fractal geometry, applied by Nature, explain that the allometric coefficient in the regression between surface area versus its metabolism is about 0.75 (West et al. 1999; West & Brown 2005). The similarly of estimates of *b* (0.66-0.83) in numerous allometric equations for various characteristics associated with metabolism (P, R, and S_{tms}) (Kamenir & Khailov 1987), the concentrations of respiratory enzymes and coenzymes and RNA (Peters 1983) may be of great importance in certain sections of the scale. One can see that the coefficients of allometries depend on the size range selected. Therefore, the most interesting are the coefficients obtained through analysis of the broadest size ranges (i.e., all living matter rather than much smaller size ranges of individual taxa): 0.75-0.80 (Peters 1983) and 0.78 for S_{tms} (Kamenir & Khailov 1987) over body mass ranges of body mass ($\Delta W\sim$ 8 to 23 orders of magnitude). Since we have the allometric coefficient b<1, the contribution of small organisms (or other small biological structures) to the production-destruction balance may be considerably greater than their contribution to biomass. For the size range of W ~ 23 orders of magnitude, the differences in specific metabolism can be high up to a factor of 10,000. Thus, the biomass of organisms should be determined the more accurately the smaller is their body size. Due to this fact, it appears logical to change over from live weight, carbon, and energy content to specific active metabolites (RNA, ATP, respiratory enzymes, and chlorophyll). Numerous methods for measuring such substances have already been developed and applied in hydrobiology.

Concluding, we can state that it is beneficial to pay more attention to aggregated metabolic surface of living organisms. It is beneficial to represent the principal groups of living matter of the World Ocean as a set of size spectra. Comparisons of several size spectra, using the active metabolic surface and allometric equations, will create better opportunities to combine data on biomass and metabolism of the main groups of oceanic organisms into an integral, non-contradictory picture that organically incorporates recent results on picoplankton biomass and production. In the complete size spectrum of living matter of the oceans, main parts of the aggregate metabolic surface, production and destruction are accounted for by comparably new groups of organisms - picoplankton (ESD < 1-2 um). Joint study of this part of the spectrum by phytoplankton biologists and bacteriologists could result in considerable revisions of the production-destruction balance and in our understanding of living matter organization in the World Ocean and its large regions.

Chapter 8

A Comparison of Eight Aquatic Ecosystems

Introduction

Study of stable patterns of community structure is among central problems of ecology. Especially important, from both a theoretical and empirical points of view, are the self-maintenance mechanisms and limits of variability of living communities (Begon et al. 1986). Numerous structural patterns, resulting from interaction of organisms with their environment, have been found by modern ecology. Several factors, that influence the structure of a given aquatic community, were studied (Odum 1971). Ecosystem size is considered by many scientists as one of the most important factors suitable for quantitative analysis. Many studies described its impact on taxonomic composition of communities (MacArthur & Wilson 1967; Begon et al. 1986) and ataxonomic size structure patterns (Schwinghamer 1981; Sprules & Goyke 1994).

Scaling is recently widely applied in many fields of sciences including biology; however, mainly at the level of living organisms. The existence of a large number of regression equations, relating important properties of living organisms to their size (ESD), body volume (V) or body weight (W) (e.g., Peters 1983; Schmidt-Nielsen 1984) provides a powerful apparatus for mathematical transformations of "ataxonomic" (Schwinghamer 1981; Steinberg et al. 1996) modeling schemes describing integral communities. Many studies relate statistical properties of communities (e.g., number of species) to the scale of the ecosystem (e.g., area of a lake or island) (MacArthur & Wilson 1967; Begon et al. 1986). Integration of two branches of ecology (taxonomic and ataxonomic models) could produce a powerful mathematical apparatus for ecological monitoring and forecasting, just as is the case of mathematical modeling in hydrodynamics and other fields of aquatic ecology.

A literature-based analysis of several traits of integral living communities was performed (Kamenir & Khailov 1987; Kamenir & Shteinman 1998) to look for typical patterns and constant estimates. Several criteria were used for the selection of the references (widely known scientific publications): analysis of systems in a maximally broad ecosystem size interval possible for our planet, comparison of natural and artificial systems, of marine and freshwater communities, of laboratory measurements and field studies data.

Material and Methods

An optimal set of aquatic communities could be composed of the world ocean, a sea, a lake, a reservoir, a pond, a mesocosm, a microcosm and a microbial vial. A series of thoroughly studied objects of aquatic ecology was compiled (Table 1) from the available literature (Kamenir & Shteinman 1998).

Well-known experimental studies, with sufficiently complete data (encompassing all fractions of living matter), were selected. Ecosystems most conforming to the described above theoretical model (IMES) were chosen, i.e., self-maintaining, closed, balanced systems with minimum nutrient addition and other anthropogenic impacts. Small, artificially created, systems were studied after a long-time uninterrupted natural succession. The values obtained at the end of the experimental period were used. To have a reasonably small data set (Table 1), eight systems were selected; inclusion of additional publications was helpful, but did not provide important corrections to the principal results.

The parameters for description of the state of the system (biotic component of the water body) were chosen on the basis of the model used (IMES). Closure of the main fluxes of the model implies that the energy influx is the main limiting factor, as its recycling coefficient is much lower than the nutrient recycling coefficient. As the energy (sunlight) flux (compensating for the energy losses) has dimensional, L^{-2} (e.g., W m^{-2}), the living matter (i.e., the totality of living beings inhabiting a unit area or the whole water area of each ecosystem) is the constraint, rather than a species or a unit volume of the environment (the water body geometric volume). Description of the living matter implies that the SS width is large and important (Kamenir 1986), hence the extreme points of the SS, i.e., the body weights of the smallest and the largest organisms (Wx, Wm) are of special interest. While study of Wx is often difficult, analysis of Wm (i.e., the body weight of the largest species) is among the traditionally measured estimates.

Table 1. Geometric parameters of analyzed aquatic ecosystems

NN	Ecosystem	Reference	log V, m^3	log Ses, m^2
1	World Ocean	Sorokin 1978	18.15	14.56
2	Black Sea	Sorokin 1985	14.73	11.62
3	Rybinsk Reservoir	Sorokin 1978	10.40	9.66
4	Lake Drivyaty	Winberg 1970	8.30	7.52
5	Outdoor Pond	Whittaker 1961	1.23	1.82
6	Indoor Pond	Whittaker 1961	-0.70	-0.40
7	Microbial Community	Taub 1969	-3.30	-2.00
8	Microbial Vial	Fishtein & Kovrov 1985	-4.30	-2.60

When the size spectrum width is large [$\log_{10}$ (Wm/Wx) >> 1)], as in many natural communities, one cannot neglect the differences in specific energy requirements of organisms of various size. The biomass (B) overemphasizes the importance of large organisms, and organism abundances (N) overemphasize the importance of small ones (Odum 1971). A power function (log Y ~ b log W, b ~ 0.75-0.80) describes this allometry (Peters 1983). The "metabolic surface" (Sm; lungs, gills, etc.) equation has a similar power coefficient b ~ 0.78

(Kamenir & Khailov 1987; Khailov & Kamenir 1998; Lebedev et al. 1989); hence, Sm is preferable to N and B when one depicts the energy dissipation according to the IMES model.

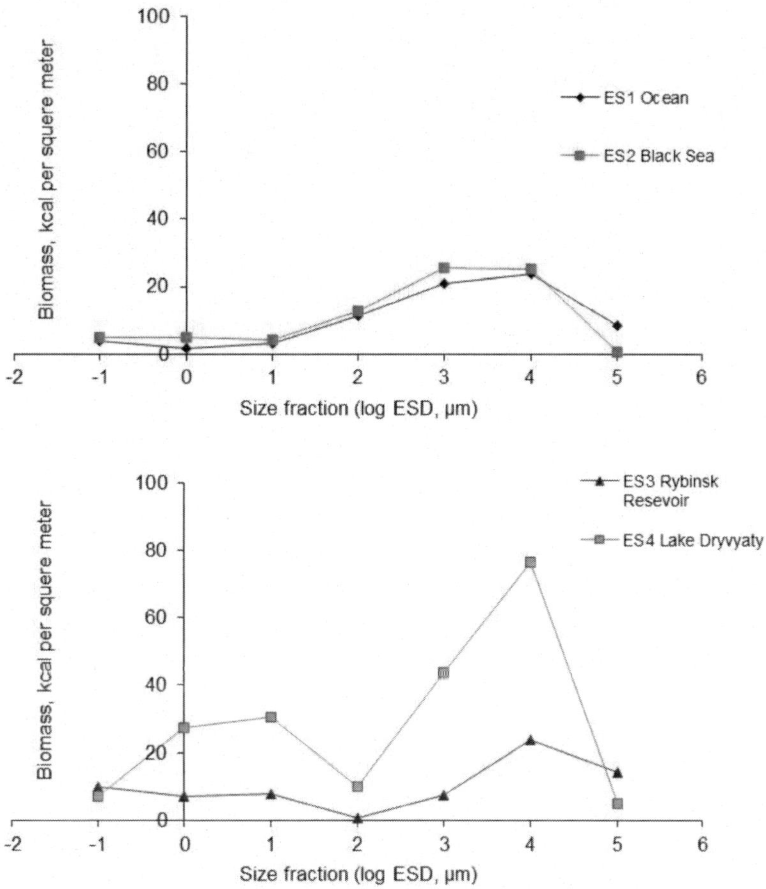

Figure 8-1. Biomass size spectra of largest aquatic ecosystems, from lakes to the ocean.

The methods of size spectra calculation and plotting were as described by Sheldon et al. (1972), Schwinghamer (1981) and Chislenko (1981). A formal quantitative parameter - size fraction 'equivalent spherical diameter' (ESD) was added to the original data to develop size spectra. All organisms were distributed into size fractions (Sieburth et al. 1978) according to their size (ESD) given by the authors or obtained from the cell volume (V), weight (W) or the biomass (B) and number of organisms (N) given: log ESD ~ 0.33 log V, where $V = W/\rho$, $W = B/N$, $\rho \sim 1$ g cm^{-3}.

Size fractions are standard increments of the organism size logarithm [$\log_{10} (ESD_2/ESD_1) = 1$]. The size fraction number corresponds to its left border, i.e., its minimum size ESD_i - ($i = \log ESD_i$, µm).

The external or 'geometric' surface of organisms was calculated using the ESD:

$St = 3.14\ ESD^2$, where the ESD for each size fraction was the mean size, i.e., the geometrical mean of ESD_i and 10 ESD_i. The allometric coefficient, accounting for the difference in the surface values of the natural organism with highly developed morph (Sm) and of the sphere of the same volume (Sg) was used (Khailov & Kamenir 1988):

Sm = St KSS, where KSS (the surface-to surface coefficient) was calculated as an allometry correction taking into account the fractal (i.e., highly dissected) geometry of geometric forms of living organisms. Summing up Sm of all organisms within all size fractions, we obtain the size spectrum of the metabolic surface. The total sum of Sm of all fractions per m^2 of the water area gives the Integral Surface Index (ISI, m^2 m^{-2}), a dimensionless parameter. The sum of all St gives the Sg integral value and Sg size spectrum of 'geometric surface.'

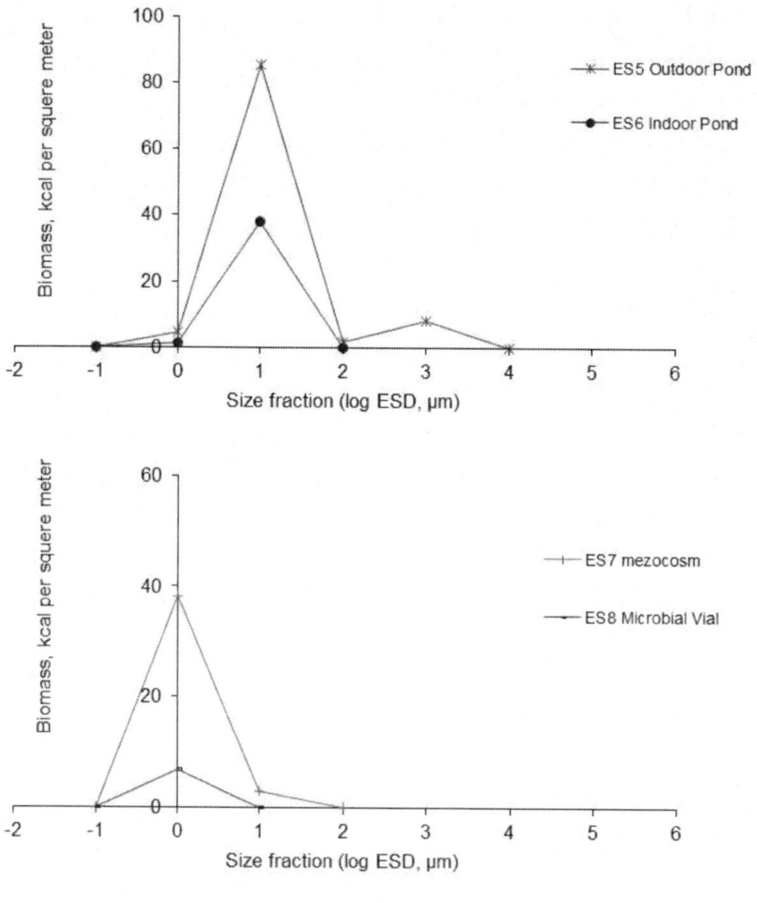

Figure 8-2. Biomass size spectra of small aquatic ecosystems (from experimental pools to microbial vials).

RESULTS

The resulting description set (Kamenir & Shteinman 1998) consisted of several estimates: the geometrical dimensions of the systems (water body volume Ves, its area Ses, mean depth h), the total biomass and total organism surface normalized to Ses (B2, ISI), and the body weight of the largest organism of the community (Wm).

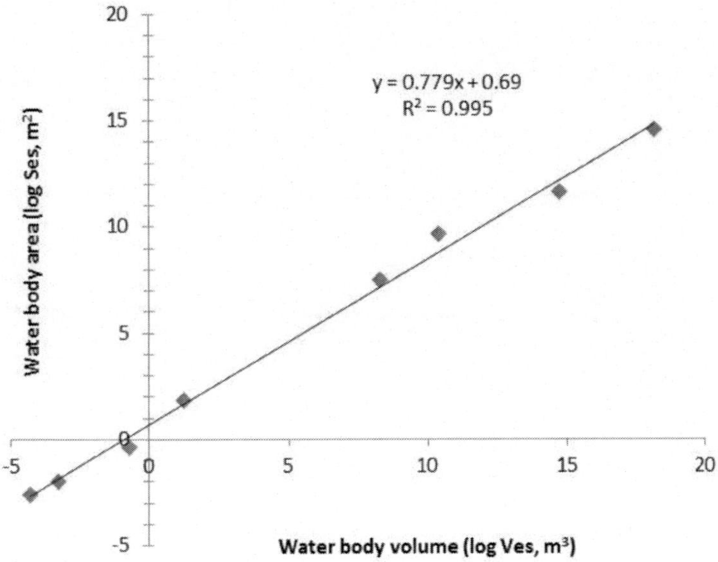

Figure 8-3. Aquatic ecosystem area (Ses) dependence on its water body volume (Ves).

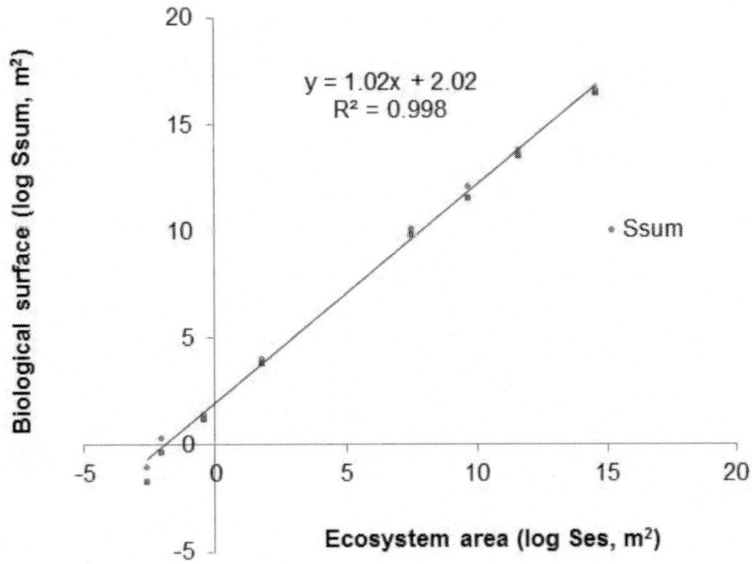

Figure 8-4. Biological surface of living organisms' dependence on ecosystem area.

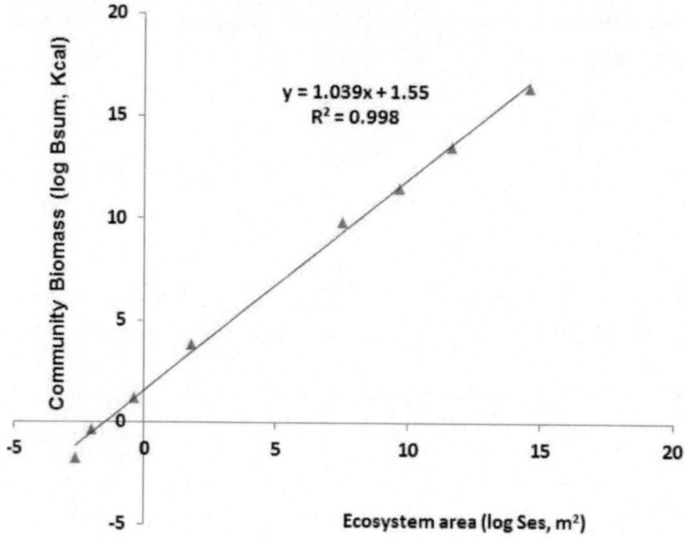

Figure 8-5. Community biomass dependence on ecosystem area.

Biomass size spectra of eight ecosystems were shown on Figures 1 and 2. Dependence of total estimates of all eight biotic communities on the ecosystem scale (Ses) was shown as graphs (Figures 3 to 6). Linear regressions of logarithmically-transformed data (Figures 3-6) produced very high coefficients of determination (r^2) and slopes close to ideal value of 1.0, i.e., results speaking in favor of our hypothesis about LIM estimates being constant if normalized to the ecosystem area (Ses). Exclusively interesting is the Ses-to-Ves (Figure 3) with slope estimate (0.779) paralleling the similar relation of living organisms SV (0.779, Kamenir & Khailov 1987).

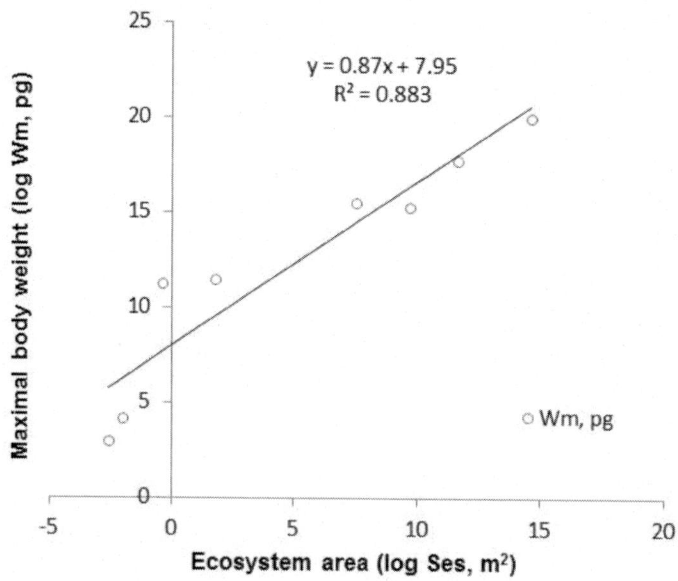

Figure 8-6. Dependence of largest organisms' body weight on ecosystem size (area).

Influence of scale parameters of an aquatic ecosystem (area, Ses, and its volume, Ves) on its morphometric and biological parameters (Kamenir & Shteinman 1998) demonstrated very high coefficients of determination of linear regression models (Figures 3-6). Especially interesting are regressions describing the total biomass of living community and total surface of LIM.

DISCUSSION

A set of water bodies of various size (some 20 orders of magnitude of water area, Ses) demonstrated a very interesting (i.e., close to ideal estimate of 1.0) slope of regression models describing dependence of integral estimates of their biotic community or LIM on ecosystem size. The estimates were: the integral mass of living organisms (Bsum), surface of their bodies (Ssum), maximal weight of the LIM particle (Wm). Such regression demonstrates that estimates normalized to Ses (i.e., B2 = Bsum/Ses, ISI = Ssum/Ses) are almost constant, i.e., not depending on the ecosystem size (Ses). Indexes of normalized total biomass (B2) were near 100 kcal m^{-2}; indexes of total surface of all organisms were about 20-200 m^2 m^{-2} (i.e., dimension-less indices). These values are close to estimates that we described previously for some other ecosystems. Size spectrum descriptions of the aquatic communities of Lakes Superior and Michigan, Scotian Shelf, Bay of Fundy and Brown Banks, including both pelagic and benthic parts (Boudreau & Dickie 1992), show almost the same biomass estimate (B2 ~ 100 g fresh weight per square meter) and size distribution. A similar estimate (100 g wet weight) was discussed previously (Schwinghamer 1981) as typical for natural aquatic systems.

Such results agree well with our estimates of the microbial biomass per unit area of the Earth (Ac, mg ATP m^{-2}), obtained earlier through bibliographic data survey and cruise works: log Ac = 1.16 ± 0.55, i.e., some 4 g carbon of microorganisms (i.e., organisms smaller than 100 um, comprising a half of the living community logarithmic size range, logWm/Wx) per m^2 (Kamenir et al. 1986; Lopukhin et al. 1990). A nearby estimate (some 4.5 gC) for integral marine planktonic communities was obtained from an extensive screening of works published in major journals (Gasol et al. 1997) for auto- and heterotrophs. Another index distinctive of low variability, i.e., integral community destruction (energy flow) per unit area of ecosystem, was discussed by Odum (1971). The right extremity of the community size spectrum, i.e., the largest organism weight (Wm) also demonstrates an approximately linear relationship with the water body area (Figure 6).

The size spectra have a pattern similar to those already described by numerous authors, i.e., several domes or bells, using terminology of Bodreau & Dickie (1992) and Sprules & Goyke (1994). The spectra are composed of bells of about the same order of magnitude, more or less symmetrical in logarithmic scales of ESD and W-axis, often, with gaps between them. The amplitudes and positions of such bells are different in different communities.

We think that such results support our idea of comparatively low variability of SS patterns and of some indices of integral aquatic communities, calculated per unit area - the living matter (B2) and metabolic surface index (ISI). The Wm regression on ecosystem size (Ses) produces a smaller determination coefficient. There are two ‘ideal ecosystems’ (Table 1, points 1 and 8) on Figure 6. These systems conform most to the ideal ecosystem (IMES) definition (i.e., describe closed spatial containers, with minimal human impact and material exchange with surrounding space): a sealed microbial ampoule and the World Ocean. One can see that almost all points in Figure 6 lie quite close to the line connecting points 1 and 8 and almost all of the points are above this line.

A possible explanation, consistent with the IMES concept described above, is that not all organisms present in the original data sets, especially in artificial experimental systems, are true members of a self-sufficient community. Some of species would become extinct if the system was hermetically sealed. For example, *Artemia salina* Linn, in a Roux flask (system 7) dies, 50 days after sealing of the system (Nixon 1969); hence, the point corresponding to this system should be moved closer to the hypothesized line connecting points 1 and 8. The same problems were seen for systems 5 and 6 described by Whittaker (1961).

The difficulties inherent in the problem of Wm estimation (the data collection and interpretation) were considered (Kamenir et al. 1998), using a study of the Lake Kinneret size spectrum. Dramatic changes in the annual and seasonal development of phytoplankton were registered in Lake Kinneret since 1994 (Berman & Shteinman 1998). For Lake Kinneret, the largest organisms are fish. Since 1993, pronounced changes in fish community composition were noted, so more recent and detailed data were obtained. Fish size spectra were calculated from acoustic data using dual beam techniques (Traynor & Ehrenberg 1979; Beurkle 1987; MacLennan & Simmonds 1992). The size spectrum of the Kinneret community biomass has a pattern similar to those known from the literature, i.e., several 'domes' or 'bells.' Analysis of detailed size spectra of fish biomass of Lake Kinneret suits well to regression models developed on the basis of previously published 8 ecosystems. As the spectra patterns hold for water bodies of considerably different scales, it seems reasonable to draw on scaling methods to model the structure and functioning of large-scale aquatic ecosystems.

Conclusion

Application of methods, developed by hydrodynamics and other fields of science studying dissipative structures, can provide a good methodology and mathematical tools for description of the aquatic community considered as a whole. Mathematical methods (size spectra and allometry) are very efficient for analysis of living communities, including the biomass distribution functions and LIM integral estimates. Size spectrum analysis of the living matter structure gave additional support for the existence of typical patterns of size spectra of integral aquatic communities. Analysis of a broad range of aquatic ecosystems, performed with the help of the IMES model, shows that for a large range of ecosystem sizes (Ses), there are parameters that change approximately linearly with the ecosystem size. An index of this type is the body weight (Wm) of the largest species. Integral community biomass (Bsum) and surface (Ssum), and community destruction also grow approximately linearly with ecosystem size; hence, such parameters, if averaged over the annual cycle period and normalized to the ecosystem size, are nearly constant for various ecosystem sizes. Very interesting is the Ses-Ves regression with the slope estimate (0.779) paralleling the similar relation of living organisms SV. A fruitful way of analysis of living communities is selection of dimensionless parameters, search for constant estimates obtained from combinations of regression slope estimates, search for logical models using such distributions and mechanisms generating them. Scaling, on the basis of typical patterns and the above-mentioned indices, can be an effective tool necessary for the development of forecasting methods in aquatic ecology.

Chapter 9

SIZE STRUCTURE STABILITY OF PHYTOPLANKTON ASSEMBLAGE IN A SUBTROPICAL LAKE

INTRODUCTION

The study of structure and stability of aquatic communities is one of the central challenges of ecology. In several studies, it has been suggested that natural aquatic communities have inherent stable size spectrum patterns (Sheldon et al. 1972; Kerr 1974; Chislenko 1981). The identification and description of such typical patterns (Schwinghamer 1981; Sprules & Goyke 1994) of size structure of natural aquatic communities provide new ways to approach the phenomenon of their stability and quantify the impact of external factors on their structure/state. The analysis of size spectra (SS) can be a valuable instrument in both empirical and theoretical studies (Sheldon et al. 1972; Kerr 1974; Platt & Denman 1978; Cavender-Bares et al. 2001). Being compact, graphic, and flexible, SS are also well adapted for mathematical transformations, comparative analysis, and generalization.

Internal mechanisms and external impacts often cause appreciable changes in the taxonomic composition of a community (Odum 1971; Begon et al. 1986). Therefore, SS and other "ataxonomic" (Schwinghamer 1981) schemes are helpful in cases of taxonomic dissimilarity, providing means to develop diagnostic means and quantitative indices of structural changes. They are suitable for description of natural systems with a large number of species, for comparative analysis of systems with different taxonomic composition; however, SS provide effective means of taxonomic interpretation (Sieburth et al. 1978; Chislenko 1981). The biomass size spectrum (BSS) has already helped establish typical patterns and develop means of quantitative comparison of BSS shape (Sheldon et al. 1972; Sprules & Goyke 1994).

Size structure of plankton communities has received much attention. Large-scale comparative studies for oceanic plankton (Sheldon et al. 1972; Rodriguez & Mullin 1986) and large groups of lakes were carried out in many countries (Sprules & Munawar 1986; Alvarez-Cobelas & Rojo 2000). A number of papers (Arin et al. 2002; Fujiki & Taguchi 2002) considered the phytoplankton size composition and its impact on primary production. From 1998, BSS analysis for the Lake Kinneret ecosystem has been conducted (Kamenir et al. 1998, 1999). This aquatic community has been intensively studied and monitored for several decades due to the great importance of the lake as the main water source in the region. From the beginning of routine monitoring of water quality parameters in Lake Kinneret in 1969 and until ca. 1993, its phytoplankton community was characterized by distinct stability, and was used in the limnological literature as one of the best known and attested examples of year-to-year similarity in abundance, distribution and composition of lake phytoplankton till the mid 1990's. Later on, the previous predictable annual pattern was lost (Zohary 2002). Numerous deviations from the established typical annual pattern coincided with a period of a pronounced reduction of the lake's water level, due to shortage of water.

Lake Kinneret is characterized with very low importance of its phytobenthos and zoobenthos. As phytoplankton produces more than 90% of the primary production in Lake Kinneret (Serruya 1978), analysis of the complete phytoplankton assemblage provides a means to study an integral autotrophic part of a natural community. The aforementioned years (1996 – 1999) with irregular water levels and phytoplankton dynamics provided a good opportunity to study phytoplankton structure (BSS) and functioning under conditions of extreme stress and fluctuations.

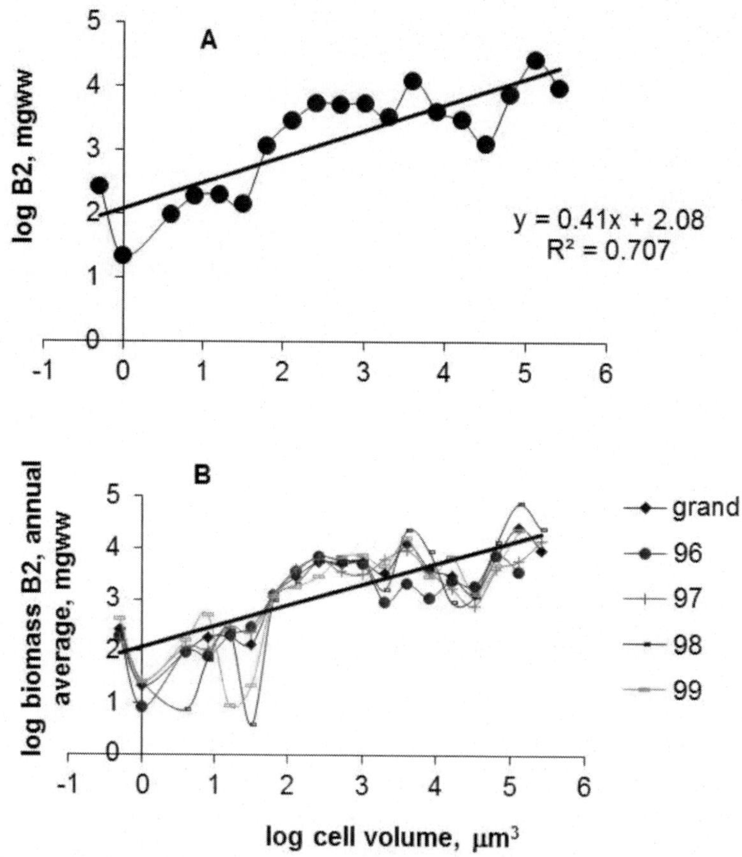

Figure 9-1. A) The general pattern of Kinneret phytoplankton depth-integrated biomass (B2) size spectrum (BSS2), showing the biomass distribution into classes of progressively increasing cell volumes. The vertical axis is the total fresh biomass (mg_{ww} m^{-2}) of organisms in a particular cell volume category. Data shown are 4-year (1996-1999) averages (n=100), plotted on a log-log scale and showing the best fit regression line. B) Annual biomass size spectra (B2) for years 1996-1999.

The aim of the next stage of studies was to analyze the stability and variability of several parameters describing the size structure of the Kinneret phytoplankton assemblage during a period of pronounced deviations from the established previously typical annual patterns.

Materials and Methods

Site Description

Lake Kinneret is situated ca. 210 m below mean sea level in the northern part of the Afro-Syrian Rift System. At full capacity, the lake surface area is some 168 km^2,, with mean and maximum depths of 26 and 44 m, respectively. Lake Kinneret is monomictic, with winter homothermal temperatures of 14-16°C, while summer maxima of the epilimnion exceed 30°C. The lake is highly productive, with annual primary production of some 600 g C m^{-2} y^{-1}, mostly produced due to the spring bloom of *Peridinium gatunense* (Hambright et al. 1997).

The lake is classified as mesotrophic to eutrophic, in summer the lake is stratified; the thermocline depth is about 15 m. For 25 years and until the mid 1990's the most salient feature of the Kinneret phytoplankton was a spring bloom of the dinoflagellate *P. gatunense,* which typically accounted for > 95% of the spring phytoplankton biomass (Pollingher 1986). The remainder of the phytoplankton biomass was mostly nanoplankton, including many minute chlorophytes, diatoms, cyanobacteria, cryptophytes and some small dinoflagellate species. These dominated the phytoplankton during summer and fall, and were readily consumed by zooplankton. Several man-induced and natural changes including the artificial lowering of the Kinneret water level and fisheries practices led to the drastically changed phytoplankton patterns since 1994 (Zohary 2002).

Phytoplankton Data Acquisition and Processing

As part of the routine monitoring program for Lake Kinneret, phytoplankton samples were collected bi-weekly from station A at the deepest part of the lake from 10-12 discrete depths throughout the water column. Lugol-preserved samples were brought to the lab for microscopic counting using the sedimentation chamber and inverted microscope technique (Utermöhl 1958), and a Zeiss M24 Axiovert inverted microscope. All phytoplankton species with individual cells greater than 2 μm diameter (cell volume, V, of 4 μm^3) were identified and counted. From the smaller cell range (picoplankton), only the relatively common colony-forming *cyanophyta* (V~ 1 μm^3), were counted.

For nanoplanktonic species (< 20 μm diameter), 10 ml subsamples were sedimented for 24 h after which all phytoplankton cells were counted at x 320. For larger species (> 20 μm diameter), 1 ml samples were sedimented and counted on the following day. Cells that were dead at the time of preservation were not counted. Phytoplankton were identified and counted according to species, and for species of highly variable cell size (like *Peridinium gatunense*), also according to size categories (e.g., small, medium, large). Phytoplankton biovolume values were calculated from species recorded abundances (density, N_i, cells ml^{-1}) and species biovolumes (V_i) approximated to simple geometrical solids (Hillebrand et al. 1999). Species biomass (B_i, mg m^{-3}) calculations were carried out assuming that all phytoplankton have a specific density of one.

Size Spectrum Analyses

Methods of biomass size spectra (BSS) calculation and plotting were used as described by Sheldon et al. (1972). All organisms counted and measured in a sample of water were distributed into size classes according to their cell volume (V_i). Size classes were standard increments of the organism size logarithm ($\Delta \log V = \log 2$), i.e., doubling of the cell volume. The Vxx notation is used throughout this paper for size classes, where xx is the class right border. The wet weight biomass of each size class was calculated using the biomass (B_i, mg m^{-3}) for each taxon and summing up the contributions from the various taxa inside each size class. The depth integrated BSS per unit area or BSS2, was calculated using data from all layers within the euphotic zone and conducting a linear interpolation between the depths. Finally, the biomasses of all classes were summed up to give the total value of the depth-integrated phytoplankton biomass shown as BSS2 in Figure 1.

Twenty five BSS2 distributions (mg_{ww} per m^2 per binary size class) were calculated each year, giving a total n=100 size spectra. The mean estimates and SD were computed for each size class (Figure 2), using all sampling dates of the studied period, but split according to year, season, or all data set. The size spectra were also analyzed as normalized BSS (NBS, Figure 2C), which were calculated via normalization of the total biomass in the i-th volume size class to the class width ($\Delta V_i \sim V_i$): $\beta_i = B_i/V_i$ (Platt & Denman 1978).

Statistical Analyses

Correlation analysis was done to find the size classes of phytoplankton most closely related to the phytoplankton assemblage biomass. Linear regression was used to estimate parameters of the SS shape. Parameters of those BSS2 linear approximations were used for comparison of intra- and inter-annual variability of the phytoplankton assemblage structure. Coefficient of variance (CV) was calculated for each size class as a measure of its seasonal variability. SPSS program, version 11.0 (Norussis 1998) was used for all statistical analyses.

Results and Discussion

During 1996 –1999, Lake Kinneret phytoplankton biomass showed high peak values (up to 512 g_{ww} m^{-2}), strong seasonal patterns, and pronounced inter-annual variability. The general pattern of Kinneret phytoplankton depth-integrated biomass (B2) size spectrum was plotted on a log-log scale to encompass very different size fraction biomass estimates at regions of small and large cells (Figure 1).

Mean BSS for these four years (n = 100) indicates the existence of two distinct peaks, or "bells" (Figure 1A), which provide the main part of variability in community biomass. These peaks have a specific taxonomic interpretation for Lake Kinneret: the highest peak, associated with size classes V32K to V128K and with average biomass for 1996 – 1999 of about 27 g_{ww} m^{-2}, was due to *P. gatunense* (cell volume 32,000 – 160,000 μm^3 with dominance of the intermediate cell sizes). The second peak, associated with the size class V4K and with average biomass of about 12 g_{ww} m^{-2} was due to *Aulacoseira granulata* (cell volume: 1100 – 6000 μm^3) (Kamenir et al. 2004).

Besides the peaks, characteristic troughs were evident in the BSS. One can distinguish here four zones (V<1, V4-64, V100-10,000 μm^3, and V64K-256K) separated by valleys of an order magnitude or more. The third bell (V100-V10,000) seems to be composed of two considerably different parts, which are V4K bell of *Aulacoseira* mentioned above, and a broad, much less vulnerable dome (V100-V1,000), which we call "nanoplankton plateau."

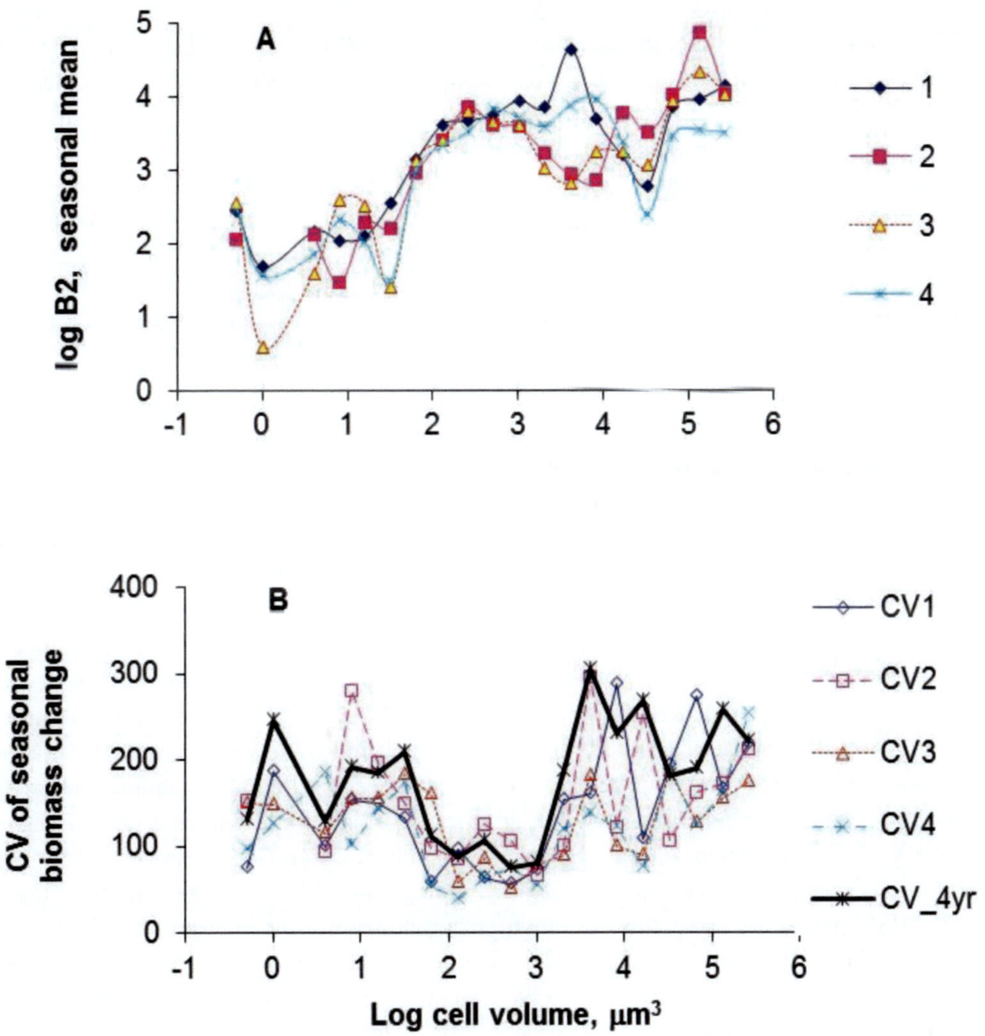

Figure 9-2. A) A Comparison of the 4-year mean average BSS (B2), for each of the seasons 1- 4 (winter to autumn). B) BSS2 seasonal variability as Coefficients of Variance (CV, i.e., SD in percent of the respective class average value) for the 4-year (1996-1999) period; n=25 for each season. The central region of low CV (logV=2 to 3) is called the nanoplankton plateau.

The variability of the BSS2 pattern helps us see the annual cycle of phytoplankton succession (Figure 2A). Here, formal division of the year into 4 seasons (3 months each) was used. As winter holomixis typically begins in December, the first part (season 1) includes months 1 and 2, and month 12 of the previous year. This way, the BSS2 change corresponds to the four seasons identified in earlier Kinneret studies: winter holomixis (season 1), spring stratification (season 2), summer stratified water column (season 3) and fall destratification (season 4). The above-mentioned periods are characterized with different composition of the lake plankton (Pollingher 1986).

The plankton composition differences are clearly seen in the formal representations (BSS2) as differences in the shape of the curve for each year or season. The various bells, composed of specific taxonomic groups, vary between years and seasons in their relative height.

Due to the pronounced temporal dynamics of phytoplankton, standard deviation (SD) was high in comparison with the mean biomass estimates, for many size classes. Therefore, instead of the commonly used Y-error bars, we present a special comparison of these estimates expressed as Coefficient of Variance (CV, i.e., SD normalized to the respective class mean value), using a special plot. Intra-annual variability of size fractions of biomass was very high (Figure 2B, coefficient of variation CV). The obtained result (Figure 2B) breaks up the studied cell size region into several parts notable with sufficiently different biomass variability (CV) (Kamenir et al. 2004).

Comparing Figures 1-2, we can conclude that BSS of integral phytoplankton community was composed of several zones (bells) separated by rather deep valleys. Three zones (V4-64, V2K-4K, and V64K-256K) were much more vulnerable than a "nanoplankton plateau" (V100-1000 μm^3), Figure 1B, 2A. Comparison of seasonal estimates averaged for 4 years (Figure 2) helps us see the size effects of temporal changes. The right extreme part of BSS2 shows a large difference between winter and summer *Peridinium* and *Aulacoseira* (V4K) biomass at the beginning and end of the winter season. Pronounced seasonal changes are also seen for small nanoplankton (V4-32), while larger cells (V64-V1K) show almost no seasonal differences.

Numerous studies consider the phenomenon of stability of size structure of natural aquatic communities, especially, plankton. Formal SS representations can complement more traditional taxonomical analyses. In the data presented (Figures 1-2), known typical patterns emerge. While BSS of the integral community is composed of several bells, it still forms a coherent system, which can be approximated by a linear regression, i.e., a simple model. The 'large algae' bell (V > 64,000 μm^3) is highly variable in time, while the central plateau (V100-1000 μm^3) is much more consistent and stable. Therefore, parameters of linear approximations and variability estimates depend strongly on the size limits set in the study.

The BSS2 pattern of Lake Kinneret phytoplankton assemblage emerges and looks 'rather stable' in spite of the drastic changes of phytoplankton biomass, annual succession pattern and taxonomic composition in 1996-1999. Statistical parameters of BSS approximation provide a scale for a quantitative estimation of this stability. Comparison of CV of specific size classes (Figure 2B) seems to be promising in this respect: It helps break up the whole cell volume range into subranges of different variability. Statistical descriptors of size classes identify two distinct zones of pronounced variability within Kinneret BSS2. This phytoplankton biomass variability is produced mainly by zones V4K and V100K, consisting of *Aulacoseira* and *Peridinium*, respectively (Figures 1-2). These algae differ considerably in their optimal requirements for many abiotic factors. Therefore, such taxonomic composition of the lake phytoplankton can produce SS variations due to changes in the abiotic environment.

Many sets of field data have several bells and gaps (Sprules & Munawar 1986; Schwinghamer 1981; Havelicek & Carpenter 2001), therefore, the linear approximation slope depends on the body mass interval considered. The slope change can be acute when the end point of BSS cuts a high bell and moves along its slope. Whereas small size is a major competitive advantage when diffusive processes control uptake at low nutrient concentrations, larger cells tend to outcompete small cells in nutrient-rich environments (Chisholm 1992; Kiorboe 1993). This scheme explains why the SS from the oligotrophic BATS station and from stratified, nutrient-poor waters in Massachusetts Bay are skewed towards smaller cells (Gin et al. 1999). Once small phytoplankton reach their maximal growth rate during nutrient enrichment, the free nutrient concentration can increase rapidly to a point where large phytoplankton can get established. So, as the total chlorophyll increases, additions are contributed from progressively larger cells (Chisholm 1992). Thus, the phytoplankton biomass growth can channel biomass to larger size classes once smaller classes receive their quota. It is widely held that marine phytoplankton communities are assembled by adding larger cells to a relatively uniform background of smaller cells. This uniform background is in fact pico and small nanoplankton. As phytoplankton communities grade from low to high biomass, there is a reduction in picoplankton, an increment in large nanoplankton, with no apparent net change in small nanoplankton (Li 2002). This uniform background of oceanic water nanoplankton bears a resemblance to the Lake Kinneret stable 'nanoplankton plateau' zone of V100-1000 (Figures 1-2).

Accumulation of the high biomass at the right extremity of SS has already been described for several types of aquatic communities in nutrient-rich environments – benthos (Schwinghamer 1991), oceanic phytoplankton (Li & Harrison 2001; Li 2002), lake microplankton (Sprules & Munawar 1986), and springs or rivers (Morin et al. 2001; Sin et al. 2000). Here, in Lake Kinneret, we see a similar phenomenon during seasonal successions of integral phytoplankton assemblage of a shallow subtropical lake. One should note the huge and variable abundance of large cells, first and foremost the largest plankton alga of Lake Kinneret (*Peridinium*). This trait is true for both seasonal and inter-annual variability (Figures 1-2).

CONCLUSION

A typical pattern of BSS2 of Lake Kinneret phytoplankton assemblage was evident. It remained quasi-stable over 4 annual cycles when the lake's water level was reduced beyond its natural levels. During these years (1996-1999) phytoplankton biomass dynamics and taxonomic composition deviated from known typical patterns. Statistical parameters of BSS2 approximation of the phytoplankton assemblage produced quantitative estimates of the algal assemblage size structure stability/variability. According to these parameters, Lake Kinneret is similar to most of the eutrophic lakes and looks a little different in comparison with stable oligotrophic systems. High values of phytoplankton biomass during nutrient inflow periods are produced mainly by large cells. Statistical descriptors of separate size classes identify a zone of stability (nanoplankton plateau) and two zones of pronounced variability within Kinneret BSS2. The plankton biomass variability is produced mainly by these two 'bloom zones' of large cells.

Chapter 10

The Long-Term Patterns of Phytoplankton Taxonomic Size-Structure

Introduction

As one can see, the BSS variation depends on the size range regions of the studied assemblage. The spectrum dynamics produce small standard deviation (SD) estimates in the central regions of the BSS-distribution, while SD estimates were much larger in both peripheral zones (i.e., of small and largest cell-volumes); hence, the spectrum stability depends on the size structure. One can suppose that SD estimates depend on the community biodiversity, i.e., on the taxonomic unit (OTUj, explained below) number and its size distribution. Such a hypothesis can be analyzed with the help of large OTUj-lists. Such a tool can be called a Taxonomic Size Spectrum (TSS).

It is a common concern that our earth's ecosystems are deteriorating due to man-induced changes (Vitousek et al. 1997; Wetzel 2001). Ecologists, therefore, seek ways to assess and quantify ecosystem stability, including means of structural similarity estimation of natural communities (Odum 1971; Begon et al. 1996; Thibault et al. 2004). Size spectrum (SS) analysis (Sheldon et al. 1972) is one of the tools available to ecologists trying to characterize whole communities. The biomass size spectrum (BSS) and normalized biomass size spectrum (NBS) are special cases of SS, where both size and biomass of the organisms in the community are expressed irrespective of their taxonomic allocation. Such 'ataxonomic' (Schwinghamer 1981) size-structure closely resembles another one, where the size-frequency distribution of species or higher taxa is plotted (Hemmingsen 1934; Hutchinson & MacArthur 1959; Chislenko 1981; Kendall et al. 1997). In this case, individual taxa (rather than individual organisms) are treated as units for accumulation by size classes. While in some important cases not all described taxonomic units are species, we refer to such size distribution of taxonomic units as traditional taxonomic size spectrum (TTSS). While BSS, and especially NBS, has gained much scientific attention in recent years, relatively few studies (e.g., Chislenko 1981; Holling 1992; Allen et al. 1999; Smith et al. 2004) addressed TTSS patterns.

Nevertheless, large-scale comparisons of animal communities have already highlighted the existence of invariant taxonomic size-frequency distributions (Roy et al., 2000; Gaston et al. 2001; Smith et al. 2004). The species size-distribution patterns observed for pelagic and benthic assemblages also seem to be rather stable, despite changes in community composition (Raffaelli et al. 2000; Havlicek & Carpenter 2001). Therefore, TTSS, a simple and flexible presentation of an important property of the community (i.e., taxonomic composition), seems to be a promising tool for quantitative assessments of whole communities. The comparison of statistical distributions of species representing community structure can be used to produce a quantitative dissimilarity index estimating the change in community composition between two points in time (Thibault et al. 2004).

In this study, we applied the TTSS approach to the phytoplankton of subtropical Lake Kinneret, Israel. We chose Lake Kinneret phytoplankton as a case study because a multi-annual phytoplankton record was available from 1969. This record included a long period of relative stability (till the mid 90s), followed by several years of strong environmental perturbations and pronounced variability in the phytoplankton annual succession, taxonomic composition, and population dynamics (Zohary 2004). The unstable period, characterized by declining water levels, abnormal blooms, and irregular seasonal phytoplankton dynamics, was already subjected to BSS analyses (Kamenir et al. 2004), which revealed a typical BSS pattern of the Kinneret phytoplankton assemblage. This pattern remained quasi-stable from year to year despite major inter-annual differences in species composition. Then the taxonomic size spectra of the most contrasting pair of years (1996 and 1998) were compared (Kamenir 2005). A broader comparison including the typical and extremely abnormal years seems to be a reasonable continuation of such analyses. While representing a singular lake, the data set of two periods, selected from the 30-year monitoring record, provided a valuable opportunity to analyze a very general and important situation of evident ecosystem destabilization. Such a data set, containing both the 'calibration' and extreme-disturbance periods, can be viewed as a specific 'acute experiment' staged by several strong and very common impacts of anthropogenic and natural origin.

The objective of our study was to compare the TTSS pattern variability of the Kinneret phytoplankton assemblage for two contrasting periods, specifically, for years of stability and years of pronounced disturbances. First, we examined if typical patterns of the assemblage structure can be established from the database of the stable period. After such patterns were found, we examined if they had survived during the extremely abnormal period in the 30-year monitoring history of the lake.

The working hypothesis was that repeating dynamics patterns are produced by communities which have some similar features in their structure. This similarity can decline or disappear during the years of pronounced changes registered for numerous biotic and abiotic parameters.

Materials and Methods

Lake Kinneret

Lake Kinneret, Israel, situated ca. 210 m below mean sea level at 32°45'N, 35°30'E (Serruya 1978), is a warm, monomictic lake with a surface area of 170 km^2, maximum depth of 44 m, and mean depth of 26 m. Man-induced changes, including the artificial lowering of the lake's water level, changes in the hydrology of the catchment area, and fishery practices, have ultimately led to drastically-changed phytoplankton patterns. For this study, we chose two subsets, referred to as the stable (1982-1985) and extreme (1996-1999) periods. The extreme period was already subjected to size-spectrum analyses (Kamenir et al. 2004). Site description, methods of the Kinneret phytoplankton community monitoring, data collection and processing, and typical patterns of phytoplankton community BSS were described in the previous chapter. Traditional methods of size spectra (BSS, NBS) creating were developed to analyze the assemblage taxonomic size structure (Kamenir et al. 2004).

Operational Taxonomic Units and Allocation to Size Classes

A mean biovolume of the individual cell of each species, or of a size category for those species counted under several size categories, was calculated from linear microscope measurements and the closest geometrical shape (Hillebrand et al. 1999). This cell volume was the parameter used for allocating a taxon to a size class. An exception was the filamentous *Aphanizomenon ovalisporum*, where the biovolume of the entire filament was used as the size unit. Since our individual taxa were not strictly species but in some cases also size categories within a species, we refer to each as an operational taxonomic unit, or OTU (Sneath & Sokal 1973).

Size classes were created by doubling the cell volume, i.e., by standard increments of the cell size logarithm. Thus, our smallest size class was $\leq 0.5\ \mu m^3$, followed by 0.5-1, 1-2, 2-4 μm^3, etc., up to the largest size class of phytoplankton in Lake Kinneret, >131,000 μm^3 (Kamenir et al. 2004). The logVxx notation is used throughout this paper for size classes, where xx is the logarithm of the right class boundary. Each class includes its right 'border' value, which is its maximal cell volume Ten size classes of the cell volume duplication correspond roughly to the 10-fold increase of the cell linear dimension D. D = 2, and 20 μm sizes delimit zones of pico/nano/microplankton (Sieburth et al. 1978).

Creating Annual Traditional Taxonomic Size Spectra

Eight size spectra were created, one for each year in the two 4-year sets (1982-1985 and 1996-1999). The total number of OTUs registered in a year (range: 62-110) was distributed to size classes (Chislenko 1981; Havlicek & Carpenter 2001) to create an annual TTSS. A fixed cell volume was used for each OTUj (Chislenko 1981; Smith et al. 2004). A TTSS was created as the frequency distribution (histogram) of the cumulative number of OTUs registered during one year to size classes. The histograms were made using the Histogram and Crosstabs procedures of the statistical package SPSS, and assigned according to the year they represented. TTSSyy notation was applied, where yy was the year, for example TTSS82 was the TTSS for 1982.

Assessing Similarity of TTSS Curves

Hierarchical cluster analysis was applied to evaluate the similarity between histograms, and to find and compare the most closely related TTSS shapes. Each histogram, i.e., a column of numbers, is interpreted as a numerical vector (Cha & Srihari 2002). The Pearson correlation (Sneath & Sokal 1973) and the Euclidean distance (ED) between the group centers (Cha & Srihari 2002) were applied to estimate the average linkage between the group centers. For the ED calculation, the input values and output proximity estimates were rescaled to the 0-1 range. The curve-fitting (linear) model was used to calculate parameters of linear regressions for XY-scatter diagrams. SPSS program version 12.0 was used for all statistical procedures.

RESULTS

Taxonomic Size Spectra

The TTSS were compared in graphic form (Figure 1). Taxonomic units (OTUs) producing each size class were listed in a special publication (Kamenir et al. 2004). All eight annual curves resemble each other by having two major peaks at the center part of the curve, at logV2.1 and at logV3.0, with lower but still elevated frequencies between those peaks. Lower frequencies with several secondary height peaks appear at both the high and low ends of the spectrum. The diverse Chlorophyta assemblage of Lake Kinneret dominates the high frequency classes at the logV1.8–logV3.3 range. Dinoflagellates dominate the right end of the curve (logV3.9 and greater. Small-celled cyanobacteria dominate the left end of the curve (logV1.2 or smaller).

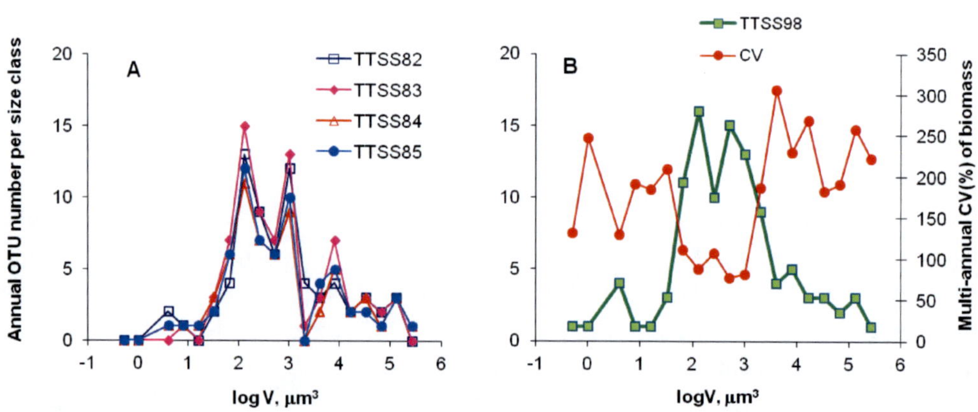

Figure 10-1. Annual taxonomic size spectrum (TTSS). A) Representative years of stability, 1982-1985. Each TTSS was created as the frequency distribution (histogram) of the cumulative number of operational taxonomic units (OTUj) registered during one year to size classes, according to the typical cell volume estimate (V, μm^3) of each OTU (see Methods). B) Coefficients of Variance (CV) from Figure 9-2 and a TTSS of a highly disturbed year (1998).

Despite the apparent similarities between the eight annual TTSS curves, deviations from the common pattern were evident. Most deviations occurred in the left (logV0.6-logV0.9) and central (logV2.1, logV2.7-logV3.9) parts of the spectra. These deviations were produced mostly by cyanobacteria (left) and several groups of chlorophytes, diatoms, cyanobacteria, cryptophytes, and small dinoflagellates (central region). The deviations were more pronounced in the curves of the unstable years (1996-1999) than in the stable years. Nevertheless, even within the unstable period, two curves (TTSS98 and TTSS99) have a large degree of overlap while they strongly differ from TTSS97 and TTSS96.

Quantification of Similarity between Curves

The results of cluster analysis were presented as a scatter diagram (Figure 2), dendrogram and its proximity measurements (Figures 3-4). The Pearson r was applied as a similarity measure (Figures 2-3). Two separate sub-clusters were seen for the stable period (1982-1985) and for the extreme period (1996-1999). The second sub-cluster was split into two pairs consisting of years 1996-1997 and 1998-1999 (Figure 4). This distinction corresponds to the pronounced change of taxonomic composition in 1998. The taxonomic composition change agrees well with the differences in the annual biomass dynamics patterns.

The two proximity estimation methods applied (the ED and the Pearson r estimators, respectively) have produced almost identical clusters, with some quantitative differences. Following the experience and argumentation of numerical taxonomy, we prefer the Pearson r measure. The graphical illustration of the TTSS comparison based on the Pearson r was shown as a linear regression model (Figure 2). Its squared value, used in the linear regression models (Figure 2), is widely applied in hydrobiology. Therefore, such an index can be helpful to scientists and other people engaged in environmental protection and monitoring.

The similarity differences reported above can be better appreciated with the help of the XY-scatter diagrams (Figure 2). Conceptually, an 'ideal' situation of stability can be described as the persistence of the same phytoplankton size structure from year to year. Thus, for any pair of years compared, each size class of one year will have the same number of OTUs as the same size class in the second year. Consequently, the pair of TTSS curves will completely overlap (Kamenir 2005). The data for this ideal pair of years can be re-plotted as an XY-scatter diagram. The data points should fall along a straight line, describing the regression, and the determination coefficient (r^2) for this line should reach the 'ideal' value of 1.0 (Sokal & Rohlf 1995). The proximity can be measured by the Pearson correlation coefficient (r) or by its squared value (r^2).

For the period 1982-1985, the similarity measures (r) between TTSS of any pair of years were very close to 1, ranging from 0.927-0.985 (Figure 3). The similarity between TTSS pairs of the years 1996-1999 was still close to 1 ($r = 0.896$-0.980). The minimum value was considerably lower than for the stable years. Thus, lower r values (Figure 3) were indicative of lower stability and the most pronounced distortions of the TTSS pattern during some abnormal years. All similarity estimates evidenced here were rather high ($r > 0.82$), including TTSS pairs separated by 10-15 years. This fact can be interpreted as the phytoplankton assemblage success in the long-term safeguarding of important structural properties.

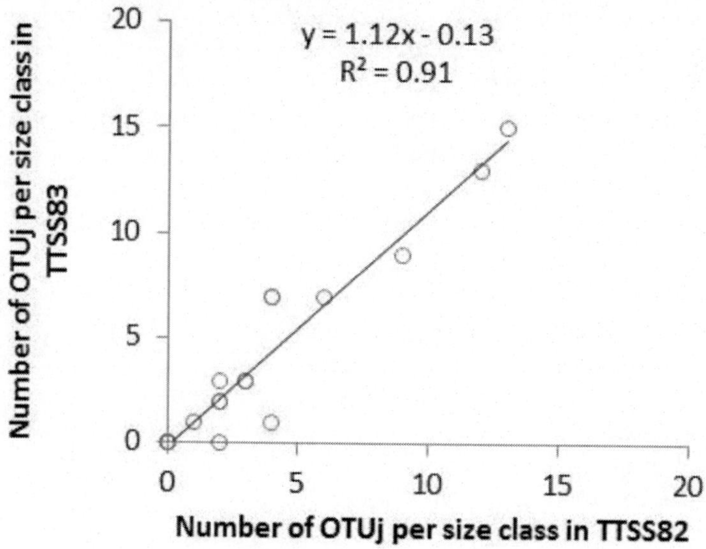

Figure 10-2. Linear regression used to illustrate the similarity measure (the Pearson r) of the TTSS shapes for pairs of years. The number of OTUs occurring in each size class in 1982 is plotted on X-axis against the number of OTUs in 1983 (Y-axis).

Proximity Matrix

	Matrix File Input							
Case	TTSS82	TTSS83	TTSS84	TTSS85	TTSS96	TTSS97	TTSS98	TTSS99
TTSS82	1.000	.957	.927	.937	.916	.908	.888	.920
TTSS83	.957	1.000	.985	.984	.901	.849	.862	.873
TTSS84	.927	.985	1.000	.984	.894	.821	.855	.848
TTSS85	.937	.984	.984	1.000	.910	.832	.855	.867
TTSS96	.916	.901	.894	.910	1.000	.949	.911	.934
TTSS97	.908	.849	.821	.832	.949	1.000	.896	.935
TTSS98	.888	.862	.855	.855	.911	.896	1.000	.980
TTSS99	.920	.873	.848	.867	.934	.935	.980	1.000

Figure 10-3. Proximity matrix for cluster analysis (Pearson r applied as a similarity measure of TTSS-vectors).

DISCUSSION

By plotting the size spectra of all species contributing to the lake's phytoplankton assemblage each year and comparing pairs of such annual spectra, we were able to show the long-term survival of the phytoplankton taxonomic size-structure patterns; and to demonstrate some sensitivity of those patterns to strong environmental perturbations. Several methods of similarity analysis (specifically, the Euclidean distance and the Pearson correlation) have provided quantitative estimates of the spectrum similarity/dissimilarity. The linear regression model and its determination coefficient should be used as the first, simple and well-known-to ecologists means (Kamenir 2005). Then, more sophisticated mathematical tools, already developed in pattern recognition sciences to compare a pair of histograms (Cha & Srihari

2002), should be applied if the phenomenon of TTSS rigidity will be proved in other ecosystems.

The TTSS, like the more popular NBS, seems to be very consistent (Gaston et al. 2001; Smith et al. 2004). Large-scale comparisons, generalizations, and theoretical discussions already exist in zoology (mainly, for terrestrial communities) (Gaston et al. 2001; Smith et al. 2004). Very close (though rather few) results have already been evidenced for planktonic communities (Havlicek & Carpenter 2001). Comprehensive studies mainly considering animals (Roy et al. 2000; Smith et al. 2004), describe very consistent and ubiquitous shapes. Our study also points to the existence of very consistent size-structure patterns. The artificial lowering of the lake's water level, changes in the hydrology of the catchment area, and fishery practices, were among the studied anthropogenic impacts. Numerous changes that took place in the phytoplankton assemblage might be due to those impacts or the subsequent adaptive responses. These changes were evidenced in the pronounced variability of the annual phytoplankton dynamics pattern, and species composition and succession. Even the biomass peak and annual average level varied greatly (Kamenir et al. 2004). In spite of all these changes, the general pattern of both NBS and log-log transformed BSS has been conserved, while definite quantitative changes of the BSS approximation parameters could be noted (Kamenir et al. 2004). The same is true for the TTSS pattern (Figures 3-4). Some basic proportions (i.e., the typical SS patterns) were safeguarded, while numerous specific details were changed. Therefore, it would be reasonable to proceed to more detailed studies.

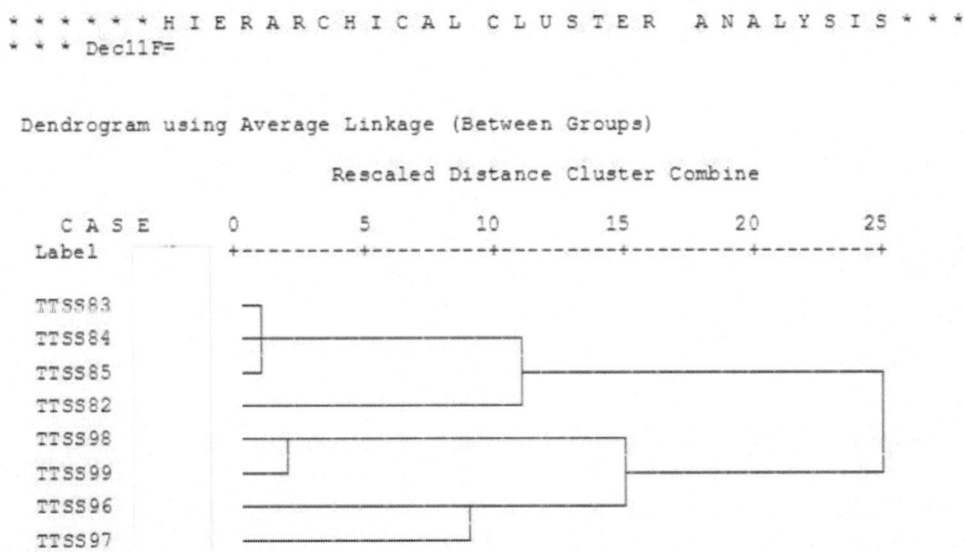

Figure 10-4. The hierarchical cluster analysis dendrogram showing similarity between size spectra considered as numerical vectors. The Pearson correlation coefficient (r) was applied as the similarity measure; r>0.8 even for pairs divided by 15 years and more. The proximity values are rescaled to the respective distances by the SPSS program.

The 'traditional' Sheldon BSS of our assemblage was definitely sensitive to some stressors (Kamenir et al. 2004). The BSS, NBS, and TTSS have different sensitivities to the disturbances and focus on different regions of the size spectrum. Therefore, their simultaneous application in such studies can lead to the establishment of both stable and variable SS details. While the initial details are valuable for predictive science, the later ones can be valuable in diagnostics. As to the TTSS, it looks very consistent, safeguarding a high similarity (i.e., the Pearson r values) even after very pronounced changes in the assemblage taxonomic composition.

While the peak positions can be fixed by some ecological mechanisms (Holling 1992; Roy et al. 2000), the size structure can be also modified by many known processes, such as eutrophication, grazing pressure, etc. (Havlicek & Carpenter 2001; Kamenir et al. 2004). For example, if a TTSS peak side shifts exist, their impact on the TTSS similarity comparisons can be considerable. The ecosystem succession, notable with the growing role of large organisms (Odum 1971), can cause a rightward peak shift and be one of the most fundamental changes of size spectra.

Body size is one of the most fundamental characteristics of a living being, strongly connected with many important aspects of its life history and ecology (Smith et al. 2004). Therefore, size groups of cells can conserve their place within a community, being pressed and supported by similar limiting factors and functional links. Conservation of functions mediated through complementary species changes must be reflected in the stable and characteristic profiles of BSS (Kerr & Dickie 2001). The stable trophic flow through the food web can depend on the development of specific adaptation mechanisms linking very different species. The apparent stability of the overall result makes us wonder about the stability of the spectrum and leads to the theories of its origin (Warwick 1984; Roy et al. 2000). The high similarity of TTSS (Figure 3) can be interpreted as a definite success of such community-level stability mechanisms. If taxonomic size distributions are a conservative property of terrestrial and aquatic communities, general ecological principles may be formulated (Havlicek & Carpenter 2001)..

The similarity measures obtained in our study are high in comparison with the similarity estimates reported by other authors (Raffaelli et al.. 2000; Smith et al. 2004), yet it is difficult to compare the values, as the above authors treated different components of the biotic community and applied different analysis methods. Havlicek & Carpenter (2001) reported similarity indices up to 0.706, though most of the values were much lower. Roy et al. (2000) and Smith et al. (2004) performed direct comparisons of body-size distributions of species using other tests such as Mann-Whitney and Kolmogorov-Smirnov two-sample test. Smith et al. (2004) also applied the linear regression and obtained the Pearson r of 0.93-0.95.

While we studied only phytoplankton, the authors cited above also included the animal components of communities. Our objective was to examine the general TTSS pattern of the whole phytoplankton assemblage, the studies mentioned above usually focused on the distribution discontinuities (clumps and gaps) or the clump shape (Holling 1992; Kendall et al. 1997) of the size spectrum. Nevertheless, the main result looks the same: the TTSS patterns seem to be stable, despite variability in environmental factors. Similarity indices of clump and gap regions between pre- and post-enrichment lakes were moderately high, even when the number of species changed considerably (Havlicek & Carpenter 2001). Our results demonstrate the stiffness of the TTSS pattern of the whole aquatic assemblage, specifically Lake Kinneret phytoplankton. This pattern withstands the species composition and even the species number variation, and is evidenced through size-frequency distributions of the OTUs comprising the assemblage.

Further comparative studies, using a wide range of ecosystems of various types, are necessary to evidence and analyze TTSS patterns under stable and extreme conditions. As TTSS analyses are based on statistical procedures, the large-scale ecosystems with higher species diversity would be valuable for such comparative studies. Such comparisons could be done without comparing long species lists, which is especially valuable in the management and monitoring of large-scale aquatic systems. TTSS could be particularly useful for the managers and environmental agencies involved in environmental monitoring and protection, because they do not require application of specialists with deep taxonomic knowledge. Just the opposite, they support the judgments produced on the bases of huge taxonomic information compressed into a quantitative index. The similarity index (e.g., Pearson r) estimates the deformation level of the whole 'object of administration' (i.e., the aquatic community or assemblage).

Conclusion

The application of a specific type of OTUj size distribution, called here traditional taxonomic size spectrum (TTSS), reveals some persistent patterns of the Lake Kinneret phytoplankton structure. Typical TTSS patterns were evident and safeguarded even during annual cycles, when many principal environmental characteristics deviated from their typical estimates and patterns established during several decades. At the same time, the TTSS reveals features that appear during perturbed years. The working hypothesis, that repeating succession patterns are produced by communities safeguarding some similar aspects of structure, was supported by quantitative statistical estimators. The general pattern of TTSS is safeguarded even during strong perturbation periods. Using modern histogram similarity techniques, TTSS can produce means of quantitative analysis of integral natural community structural variability. Such means are acutely needed by environmental management, monitoring, and theoretical ecology.

Chapter 11

Comparison of Taxonomic Size-Structure of Several Aquatic Ecosystems

Introduction

It is well known that numerous natural ecosystems are deteriorating due to man-induced activity (Vitousek et al. 1997; Wetzel 2001). Ecologists, therefore, try to assess and quantify ecosystem stability of natural communities (Odum 1971; Thibault et al. 2004). The biomass size spectrum (BSS) and normalized biomass size spectrum (NBS) are special cases of size spectra, where both size and biomass of the organisms in the community are expressed irrespective of their taxonomic allocation. Such 'ataxonomic' size-spectra resemble another type of size distributions, where the size-frequency distribution of species or higher taxa is plotted (Hemmingsen 1934; Hutchinson and MacArthur 1959; Chislenko 1981; Kendall et al. 1997). In this case, individual taxa are treated as units for accumulation by size classes. In many important cases not all described taxonomic units are species; therefore, we consider operation taxonomic units (OTUj) and refer to such size distribution of taxonomic units as traditional taxonomic size spectrum (TTSS). Well-known BSS and NBS have gained much scientific attention in recent years, while relatively few studies (e.g., Holling 1992; Allen et al. 1999; Smith et al. 2004) addressed TTSS patterns. Nevertheless, large-scale comparisons of animal communities have also shown the existence of invariant taxonomic size-frequency distributions (Roy et al. 2000; Gaston et al. 2001; Smith et al. 2004). The species size-distribution patterns observed for pelagic and benthic assemblages seem to be rather stable, in spite of changes in community composition (Raffaelli et al. 2000; Havlicek & Carpenter 2001). Therefore, TTSS, a simple and flexible presentation of an important property of the integral community (i.e., its taxonomic composition), seems to be an efficient tool for quantitative assessments of large communities. The comparison of statistical distributions of species, representing the community structure, can be used for quantitative similirity/dissimilarity indices, to analyze the change in community composition as a process, to analyze a specific time change (Thibault et al. 2004).

We had already applied the TTSS approach to the phytoplankton of subtropical Lake Kinneret, Israel (Kamenir et al. 2006, 2008b). Thus, it seems that the phytoplankton TTSS pattern is likely to be supported by some ecological mechanisms of a very general nature and demands further studies considering specific objects and situations of general ecological interest (Kamenir et al. 2008b). The high self-similarity of the TTSS general pattern during long-term periods can be interpreted as the strength of the ecological community self-maintenance. The role of specific anthropogenic and natural impacts – such as aquatic system size, depth, and trophic status change – seems to be very important, especially during times of global change. Therefore, we continued studies on phytoplankton TTSS, selecting a small set of considerably different aquatic systems (a mountain lake in Europe), an oligotrophic lake from a very distant location (the western side of American continent), and a marine system (the English Channel). We applied the TTSS approach to analyze structural changes of the phytoplankton assemblage, considering general traits and typical patterns, trying to override the differences in specific details (environment, ecological monitoring, data collection, and processing methods) of local and historical nature (Kamenir et al. 2006, 2008b, and 2010; Kamenir & Morabito 2009).

The Objects, Objectives, and Hypotheses

We chose a deep subalpine Lake Maggiore (LM) in Italy as a case study because a multi-annual phytoplankton record was available, and a clear and thoroughly documented oligotrophication process took place there (e.g., Salmaso et al. 2007).

Lake Tahoe (LT) was selected as it has experienced several major changes over recent decades, caused by watershed alterations, including an overall decline in transparency. Tahoe is a large, deep, oligotrophic mountain lake, located at an altitude of almost 2 km in the Sierra Nevada, USA; hence, we expected to find strong contrasts in many traits of the three lakes (LK, LM, and LT). LK means Lake Kinneret.

Our next step was a comparison of fresh-water and marine phytoplankton. The Plymouth marine laboratory (PML) has a long-term time-series station in a coastal temperate environment in the western English Channel, where microplankton samples have been collected on approximately a weekly basis since 1992.

All four aquatic systems were monitored over multi-annual time periods (20 years and more).

The objectives of study were:

- for LM, we tried to examine the TTSS pattern change of the phytoplankton assemblage during a well-documented oligotrophication process. We tried to establish specific trends and break down the total 22-year range into several periods of more homogeneous (i.e., self-similar) patterns.
- For LT, the goal was to compare the phytoplankton TTSS of two geographically distant and limnologically different (a cold oligotrophic and a warm meso-eutrophic) lakes.
- For PML, the general aim was to compare the phytoplankton TTSS of two geographically distant and hydrographically different water bodies (a sea and a fresh-water lake).

The working hypotheses were:

- repeating annual succession patterns are produced by communities that preserve some reliable structural features. We examined if typical patterns of the assemblage structure can be established.
- This reliability can decline during the years of pronounced changes of the ecosystem state. Different typical patterns may exist during the stable and transitory stages.
- The general patterns of phytoplankton TTSS are consistent across an inter-annual time series within an ecosystem and have important common properties across aquatic ecosystems, while some fine details of the TTSS patterns are determined by ecosystem structure and can vary from one system to another.

MATERIAL AND METHODS

Monitoring Sites

Lake Kinneret. LK (Israel) is monomictic, situated some 210 m below mean sea level, in a subtropical rift valley. It is meso-eutrophic, warm (surface layer temperatures up to 30ºC), surface area of 170 km^2, and maximal depth of 43 m. Hydrology, data collection and processing for LK were described in detail above (see chapters 8-10).

Long-term oligotrophication process of Lake Maggiore. Lake Maggiore, the second largest subalpine lake in Italy (lake area of 213 km^2), is situated at 193 m above mean sea level. It has a maximum depth of 370 m and an average depth of 178 m (Salmaso et al. 2007). It is oligotrophic by nature, as shown by early limnological studies (Baldi 1949) and by analysis of the sedimentary pigments (Guilizzoni et al. 1983). The eutrophication began in the sixties - the phosphorus concentration in the lake water began to rise, followed then by an increase in phytoplankton abundance, biovolume, and primary production (Ravera & Vollenweider 1968; Morabito & Pugnetti 2000).

In the late seventies, the lake reached a trophic state close to eutrophy (Mosello & Ruggiu 1985). Later, the P loads have been gradually reduced by various means. As a result, the TP values have gradually decreased to some 10 μg L^{-1} (Ruggiu et al. 1998). The long-term oligotrophication of Lago Maggiore is documented in many papers. Strong emphasis was put on the apparent resilience of the plankton communities during phosphorus concentration decline (de Bernardi et al. 1988). From 1987-88, major biological changes were revealed in the phytoplankton (Manca et al. 1992). A notable decrease in average cell size and increased importance of small size algae were notable (Ruggiu et al. 1998).

Ecological perturbations at Lake Tahoe. LT is a large, deep, oligotrophic mountain lake, located at an altitude of 1898 m in the Sierra Nevada mountain range, USA. The surface area of the lake is 501 km^2 and the watershed is 812 km^2. Maximum depth is 501 m, and mean depth is 313 m. The lake is warm, monomictic and free of ice the entire year. Thermal stratification typically begins to intensify during February–April; in August the thermocline is at some 30 m below the surface (Goldman et al. 1989). LT has high transparency and a deep-water chlorophyll maximum. LT has been monitored continuously since the late 1960s (Goldman et al. 1989). The lake has experienced several major changes over recent decades caused by watershed alterations (Jassby et al. 2003), a near-steady increase in primary productivity (Goldman et al. 1989), a shift in nutrient limitation from phosphorus and nitrogen co-limitation to nitrogen limitation (Goldman 1988). Biologic changes included the introduction of several invertebrate and fish species (Vander Zanden et al. 2003), and the subsequent disappearance of zooplankton species. Two periods, separated by a long period of ecological perturbations, specifically 1983 –1986 and 2002–2005, were selected from the LT database for comparative studies.

The Plymouth marine site at station L4. PML is a long-term time-series station (http://www.westernchannelobservatory.org.uk/l4/) situated in a coastal temperate environment in the western English Channel, 10 km southwest of Plymouth, 50°15' N, 4°13'W (51-m deep). The physical and chemical environment have been monitored and documented in a number of publications (e.g., Rodriguez et al. 2000). Temperatures are 8ºC-16ºC. The thermocline gradually deepens during the summer, and then weakens. Stratification is reestablished in August. Convectional cooling and autumn storms in late September result in the deepening of the surface mixed layer. Complete vertical mixing continues through winter. Nitrate concentrations are up to 7 μmol (in the winter) and decline to <0.4 μmol L^{-1} in April. The spring bloom usually occurs in April.

Phytoplankton Data Acquisition and Processing

Lake Maggiore. Water samples for phytoplankton analysis were collected fortnightly (monthly during the cold months - November to January), at the station of Ghiffa, at the deepest point of the lake. A bottle, for an integrated sample in the 0-20 m water layer, was used (R. Bertoni, 1996; Patent. Num. MI 96/A 000121). The device allows a continuous sampling from the water column, taking a fixed volume of water at each meter. Phytoplankton was determined in subsamples of 10 ml preserved in acetic Lugol's solution. Algal cells were identified and measured on a Zeiss Axiovert 10 microscope, following Lund et al. (1958), until 400 cells of the most important species were counted. Algae >2 μm (GALD) were classified and included into our analysis. Algae up to some 1 μm in GALD were classified among the small Chroococcales when in colonies; single cells were considered as unidentified picoplankton; picoplankton was not included in the current analysis.

For determination of the phytoplankton species the standard systematic literature was used. Oscillatoriales were determined also following Anagnostidis & Komárek (1988). Water for chemical analyses was collected monthly; nutrient concentrations were determined at the chemical laboratory of the Istituto per lo Studio degli Ecosistemi (Pallanza), following Mosello & Ruggiu (1985). All procedures of data sampling and processing were very similar to protocols followed up at LK (Zohary 2004; Kamenir et al. 2006).

Lake Tahoe. The phytoplankton enumeration and taxonomic identification in LT was comparable to that used for Lake Kinneret (Zohary 2004). As part of the routine monitoring, composite phytoplankton samples from the upper 100-m water column were collected at mid-lake and near-shore stations (for station location, see Jassby et al. 2003) and fixed with Lugol's solution. The composite sample was composed of volume weighted aliquots, collected from several depths at each station, based on the hypsographic curve of the lake. After sedimentation of the100-ml water sample for 72 h (Lund et al. 1958), cells were identified, measured and counted at the species level with an inverted microscope at 150×, 300×, and 600× magnification, including algae with cell volume > 4 μm^3. In each sample a minimum of 400 cells was counted and linear dimensions were measured for dominant species. The cell volume was estimated by assigning different geometric shapes (Hillebrand et al. 1999). A mean cell volume was assigned for each phytoplankton species based on average cell-size measurements over the sampling period 1983 – 2005. The estimated cell volume is the parameter used for allocating a taxon to a size class, even for colonial species, as was also done in the case of LK (Zohary 2004). Since our individual taxa were not always strictly species, we refer to each as an operational taxonomic unit, or OTUj (Sneath & Sokal 1973). Size classes were created by doubling the cell volume (V), i.e., by standard increments of the cell size logarithm. The Vxx notation is used for size classes, where xx is the logarithm of the class right boundary (Kamenir et al. 2006).

*Plymouth marine station (PML)***.** Microplankton samples have been collected on approximately a weekly basis since 1992. Paired samples are collected from a depth of 10 m and preserved with 2% Lugol's iodine solution and 4% buffered formaldehyde. Sample amounts from 10 to 100 ml (depending on cell density), are settled for >48 h and identified to species level (when possible) and enumerated using an inverted microscope. The dimensions of individual species are measured, and cell volumes calculated (Kovala & Larrance 1966). All phytoplankton species with individual cells > 2 um are described in the monitoring database (L4).

The traditional taxonomic size spectrum (TTSS) development was applied as described by (Kamenir et al. 2006). A mean biovolume (V, μm^3) of the individual cells of each species, or size category for species counted under several size categories, was the parameter used for assigning a taxon to a size class. The Vxx notation is used throughout this paper for size classes. The average, minimal, and maximal estimates of the annual species number were obtained for each size class from all TTSS [excluding the year 1992].

The TTSS development methods are described above (Kamenir et al. 2006) for all four aquatic systems. Since not all our taxa were strictly identified species, we refer to each as an operational taxonomic unit, or OTUj (Sneath & Sokal 1973). Treating a histogram (TTSS) as a numerical vector, we estimate the similarity measure for each pair of spectra with the help of hierarchical cluster analysis (Cha & Srihari 2002). Numerous TTSS were created for each ecosystem, one for each year. A TTSS was created as the frequency distribution (histogram) of the total number of OTUj-s registered during one year to size classes. The histograms were built using the Histogram and Crosstabs procedures of the statistical package SPSS, and assigned according to the year they represented. LLttXX notation was applied for ecosystem-code LL, and XX was the year (Kamenir et al. 2006). The dynamics of each size fraction was described by its OTUj number registered and average cell volume (Lake Maggiore) per year. SPSS program, version 14.0, was used for all statistical procedures.

RESULTS

Lake Maggiore. Throughout the 22 years of analysis, the phytoplankton assemblages of L. Maggiore were made up by > 200 taxonomic units. The species number increase is clearly seen in the long-term dynamics curve (Figure 1). The observed phytoplankton assemblage evolution was connected with an overall decrease in phytoplankton average cell volume. This reduction exhibited a very clear pattern during the years 1986-1997 (Figure 1). Two main trends caused this volume decrease: the size changes occurring inside each algal group and the overall reduction of the assemblage mean size due to the addition of numerous medium-and-small species. Species number change in the most important size fractions had obviously non-linear dynamics. This dynamics was paralleled by dynamics of taxonomic size spectra. All annual TTSSs, compared in graphic form, show variable shapes; however, the same general pattern is seen while comparing TTSS shapes (Figure 2). This pattern represents several modes of more-or-less the same height within the size (logV) range of 1 to 4, with much lower peaks in both peripheral zones. We can see several peaks divided by rather deep troughs. The number of peaks and their positions slightly vary at times. With the help of hierarchical cluster analysis, this set of shapes was divided into several sub-clusters or periods characterized with higher homogeneity (Figures 2 and 3). One can see almost the same shape of four central peaks (region logV = 1-4) with much lower peripheral bells at logV = 0.0-0.9 and logV = 4.2-5.1 (Figure 2).

The highest level (r = 0.966) of the similarity measure (Pearson r) in pairs of annual TTSS (Figure 3) was close to the respective estimates found for other lakes (Kamenir et al. 2006, 2008b). Its minimal level (r = 0.669), described a pair of the years (1989 and 2003) at the beginning and the end of the studied process. Formally, considering Pearson r > 0.9 as 'high' similarity levels, we can distinguish several 'homogenous periods' – several pairs of years (1984-1985 and 1988-1989; 1990-1991) and following that, three long time-intervals (1992-1995, 1996-2000, and 2002-2004) (Figure 3). Some years exhibit 'intermediate' TTSS shapes, having high similarity with their neighbors both on the left and right

Consequently, we can see the transformation process, divided into several stages (Figures 2 and 3). This process ended in 2005 with a 'return' to the initial shape (1984), but with a much upgraded height, i.e., the higher number of OTUj (Figure 1).

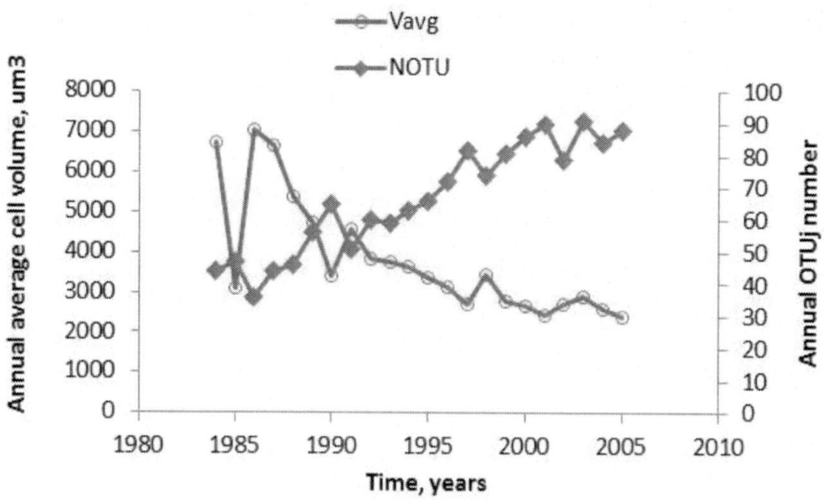

Figure 11-1. Long-term dynamics (1984-2005) of the annually registered OTUj number and of the annual average cell volume (V, μm^3) produced by the annual taxon list of Lake Maggiore.

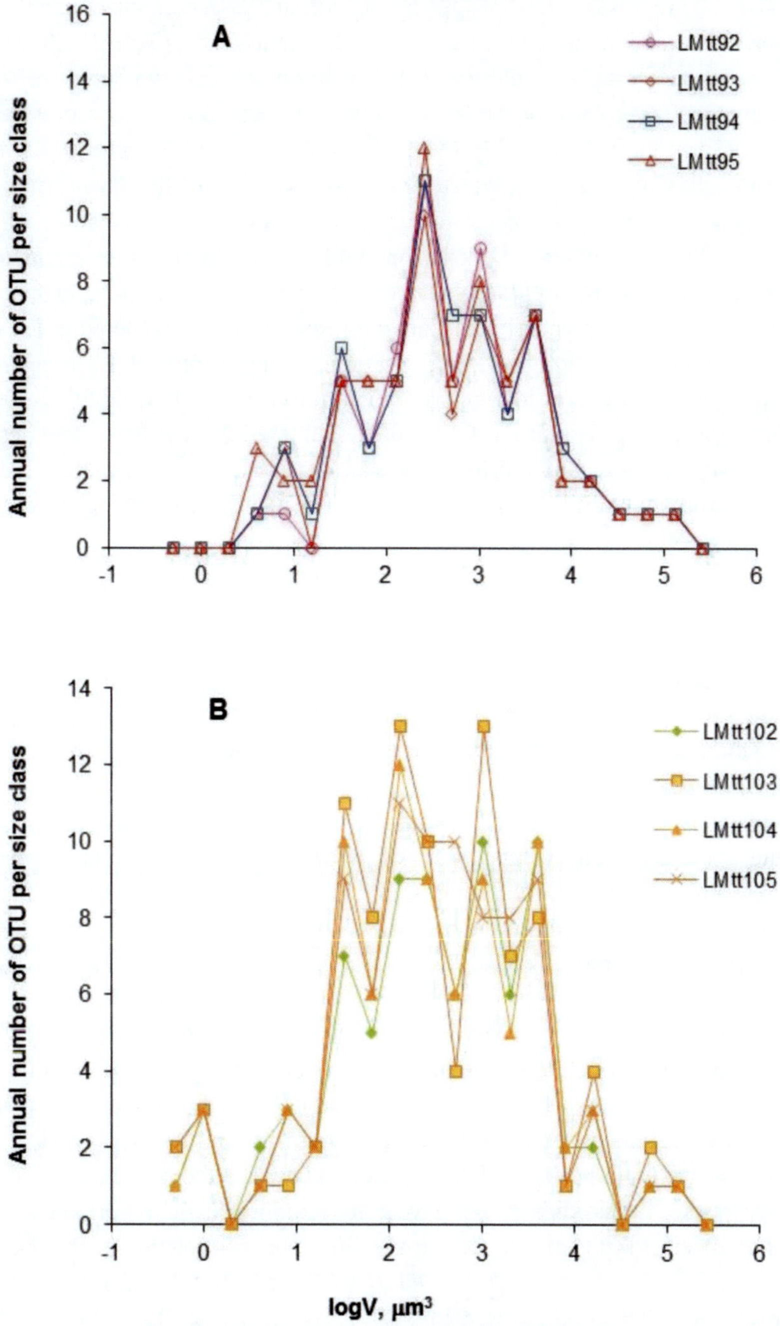

Figure 11-2. A comparison of the L. Maggiore phytoplankton TTSS-shape for two periods representing the period of the most noticeable phytoplankton structure transformation – 1992-1995 (A) and years following it (2002-2005).

Two different shapes of the phytoplankton taxonomic size structure can be distinguished, characterizing the transitory period and the 'less dynamic years' preceding and following it (Figure 2).

Lake Tahoe. The representations of LT phytoplankton TTSS demonstrate the existence of almost the same general annual pattern for each of the eight years (Figure 4). Several distinct peaks can be seen each year, although the exact position and relative height of some peaks varied between years. One can note zones at logV = 2.1 and 2.7-3.9 as more variable than other zones. These properties resemble the Lake Kinneret TTSS structure (Kamenir et al. 2006). Exceptionally good similarity was noted for the left side of the spectrum, log V < 2.7 (i.e., cell volume < 300 μm^3). The species contributing to these size classes were different for the two lakes. In Lake Kinneret the most abundant species in this size range were small coccoid cyanobacteria and small chlorophytes, while in LT, the most notable species of this size range were diatoms and cryptophyta. Many new species were registered in LK during the disturbed (extreme) period (Zohary 2004). Taxonomic composition of LT phytoplankton also evidently was different while considering the two periods (1983-1986 and 2002-2005). Many species common during the first period were not encountered during the second period, while many species have appeared. Such differences were especially notable in size fractions with small species number, but similar, though less evident situations, were registered practically in all other fractions.

The cluster analysis has demonstrated high similarity levels of TTSS in Lake Tahoe (Kamenir et al. 2008b) even for spectra >20 years apart from each other. Another cluster analysis, was performed for the 16 annual TTSS vectors, eight described above (Lake Tahoe) and eight developed earlier (Kamenir et al. 2006) for Lake Kinneret. The resulting dendrogram demonstrated rather high similarity for all 16 TTSS (with a similarity estimator, Pearson correlation coefficient, $r > 0.68$), but distinguished between the two lakes (LK and LT) that formed separate sub-clusters.

Marine phytoplankton (PML). The 14 annual TTSS curves of station L4 phytoplankton were built and applied for cluster analysis. All TTSS patterns had similar general shapes, showing peaks and troughs at the same size classes (Figure 6). Concomitantly, some variations were notable, especially in the regions of the dominant peaks (V2.1 and V4.2) and at the main trough between them (V3.0).

While such detailed comparisons of specific years and size fractions seem to be helpful in ecosystem monitoring, the inter-ecosystem comparison provides several common characteristics and very distinctive quantitative differences. The general pattern of station L4 (Figure 5A) resembles the respective pattern of subtropical Lake Kinneret (Figure 5B), described previously (Kamenir et al. 2006). Both patterns are almost symmetrical in the log (V) scale and have two main peaks, with a deep trough between them. In both ecosystems, the height of these peaks varies slightly from year to year, while their horizontal position is unchanged. The most striking difference between the two ecosystems is the second major peak position, which is V3.0 for LK compared to V4.2 for station L4. As the first peak (i.e., small-size algae) takes the same position in both ecosystems (V2.1), the inter-peak range is some two times broader in the L4 spectra. This first peak (V2.1) coincidence is very interesting and requires an explanation, as we consider two very different aquatic ecosystems. This peak takes the same position also in LT (Figure 4) phytoplankton TTSS (Kamenir et al. 2008b). The total cell-size range is much shorter for Lake Kinneret, as its largest phytoplankton species (*Peridinium gatunense*) has logV ~ 5 (Zohary 2004), while the marine station L4 included species (*Rhizosolenia* spp. and *Coscinodiscus* spp.) of logV ~ 6-7 (Figure 6), e.g., *Coscinodiscus wailesii* .

	LM84	LM85	LM86	LM87	LM88	LM89	LM90	LM91	LM92	LM93	LM94	LM95	LM96	LM97	LM98	LM99	LM100	LM101	LM102	LM103	LM104	LM105
LM84		0.943	0.815	0.890	0.918	0.933	0.865	0.835	0.879	0.845	0.899	0.834	0.697	0.727	0.788	0.723	0.836	0.887	0.849	0.712	0.812	0.884
LM85	0.943		0.859	0.898	0.931	0.919	0.904	0.832	0.892	0.858	0.875	0.877	0.757	0.815	0.841	0.831	0.860	0.905	0.873	0.752	0.811	0.869
LM86	0.815	0.859		0.921	0.867	0.869	0.835	0.841	0.883	0.816	0.824	0.834	0.739	0.686	0.799	0.782	0.770	0.788	0.805	0.678	0.680	0.688
LM87	0.890	0.898	0.921		0.944	0.934	0.907	0.885	0.909	0.838	0.860	0.840	0.718	0.694	0.787	0.776	0.804	0.811	0.800	0.691	0.724	0.735
LM88	0.918	0.931	0.867	0.944		0.885	0.914	0.880	0.882	0.842	0.830	0.815	0.687	0.704	0.760	0.787	0.805	0.837	0.854	0.752	0.804	0.812
LM89	0.933	0.919	0.869	0.934	0.885		0.912	0.847	0.913	0.859	0.931	0.884	0.781	0.786	0.894	0.812	0.895	0.893	0.797	0.669	0.743	0.805
LM90	0.865	0.904	0.835	0.907	0.914	0.912		0.915	0.954	0.920	0.896	0.921	0.825	0.826	0.869	0.879	0.923	0.911	0.871	0.804	0.828	0.807
LM91	0.835	0.832	0.841	0.885	0.880	0.847	0.915		0.961	0.916	0.894	0.891	0.800	0.746	0.808	0.810	0.857	0.885	0.901	0.902	0.891	0.829
LM92	0.879	0.892	0.883	0.909	0.882	0.913	0.954	0.961		0.957	0.961	0.957	0.862	0.816	0.877	0.844	0.918	0.930	0.909	0.830	0.849	0.838
LM93	0.845	0.858	0.816	0.838	0.842	0.859	0.920	0.916	0.957		0.949	0.966	0.924	0.893	0.900	0.899	0.955	0.916	0.908	0.826	0.854	0.855
LM94	0.899	0.875	0.824	0.860	0.830	0.931	0.896	0.894	0.961	0.949		0.948	0.862	0.829	0.909	0.817	0.932	0.929	0.881	0.751	0.832	0.864
LM95	0.834	0.877	0.834	0.840	0.815	0.884	0.921	0.891	0.957	0.966	0.948		0.958	0.919	0.929	0.895	0.963	0.923	0.882	0.791	0.806	0.822
LM96	0.697	0.757	0.739	0.718	0.687	0.781	0.825	0.800	0.862	0.924	0.862	0.958		0.948	0.920	0.906	0.938	0.853	0.783	0.723	0.700	0.728
LM97	0.727	0.815	0.686	0.694	0.704	0.786	0.826	0.746	0.816	0.893	0.829	0.919	0.948		0.930	0.946	0.933	0.854	0.762	0.720	0.707	0.764
LM98	0.788	0.841	0.799	0.787	0.760	0.894	0.869	0.808	0.877	0.900	0.909	0.929	0.920	0.930		0.943	0.958	0.910	0.809	0.727	0.754	0.802
LM99	0.723	0.831	0.782	0.776	0.787	0.812	0.879	0.810	0.844	0.899	0.817	0.895	0.906	0.946	0.943		0.932	0.865	0.808	0.776	0.758	0.761
LM100	0.836	0.860	0.770	0.804	0.805	0.895	0.923	0.857	0.918	0.955	0.932	0.963	0.938	0.933	0.958	0.932		0.945	0.869	0.783	0.824	0.851
LM101	0.887	0.905	0.788	0.811	0.837	0.893	0.911	0.885	0.930	0.916	0.929	0.923	0.853	0.854	0.910	0.865	0.945		0.934	0.845	0.902	0.923
LM102	0.849	0.873	0.805	0.800	0.854	0.797	0.871	0.901	0.909	0.908	0.881	0.882	0.783	0.762	0.809	0.808	0.869	0.934		0.916	0.961	0.937
LM103	0.712	0.752	0.678	0.691	0.752	0.669	0.804	0.902	0.830	0.826	0.751	0.791	0.723	0.720	0.727	0.776	0.783	0.845	0.916		0.945	0.880
LM104	0.812	0.811	0.680	0.724	0.804	0.743	0.828	0.891	0.849	0.854	0.832	0.806	0.700	0.707	0.754	0.758	0.824	0.902	0.961	0.945		0.943
LM105	0.884	0.869	0.688	0.735	0.812	0.805	0.807	0.829	0.838	0.855	0.864	0.822	0.728	0.764	0.802	0.761	0.851	0.923	0.937	0.880	0.943	

Figure 11-3. The hierarchical cluster analysis (Proximity Matrix) showing similarity between size spectra considered as numerical vectors. The Pearson correlation coefficient (r) was applied as the similarity measure; $r > 0.8$ even for pairs divided by 15 years and more. LMxx means year xx for Lake Maggiore.

The cluster analysis results produced the TTSS similarity estimates (Pearson r) for all pairs of years ($r = 0.861$-0.992). This range was quite comparable to that obtained previously (Kamenir et al. 2006, 2008b) for all 3 lakes considered above (LK, LM, and LT). The inter-ecosystem (L4 and LK) pairs had much lower similarity levels ($r = 0.317$-0.578). While the peak heights changed from year to year, their positions were invariant.

The TTSS multi-annual consistency was demonstrated for a marine ecosystem. The evident and permanent differences in the peak positions between the marine and fresh-water ecosystems (PML and LK) open a way for additional analyses which can be helpful for the development of theoretical ecology and ecological monitoring.

DISCUSSION

Lake Maggiore. Long-term analysis of taxonomic structure change during an apparent lake oligotrophication process (Ruggiu et al. 1998) shows many interesting characteristics of integral phytoplankton assemblage adaptation. We can see the transformation process of the lake phytoplankton taxonomic structure - abrupt, for some size fractions, and almost linear for the total species number (Figure 1) dynamics coming to a new situation that reproduced itself after several years. This situation is clearly seen when comparing several short periods (Figure 2).

The most drastic restructuring is evident within the TTSS central region, especially for the notorious fraction V2.4. The most prominent overall changes describe total taxonomic diversity growth and the general decline in the average species size (Figure 1). Another prominent property (seen with the help of TTSS and cluster analysis) is the existence of several clearly different stages (Figures 2-3), where the first and last have almost the same general pattern of three equal height peaks; the intermediate transitional stage shows an explicit domination of one central peak (Figure 2). During all periods, we can see peaks at positions (logV) of 2.4, 3.0 and 3.6. Only in the last period (2002-2005) one of the peaks (logV = 2.4) moved one class leftward while logV = 2.4 went down. This process is clearly seen via TTSS dynamics (Figure 2).

Change in size structure is influenced by many phytoplankton groups. The clear cell volume decrease of dinoflagellates was due to the appearance of small *Gymnodinium* species. It is very plausible that the rearrangement of the phytoplankton assemblage was a consequence of the ecosystem changes related to nutrient reduction (Kamenir et al. 2009). This reduction could have given a growing advantage to species with a higher SV-ratio, being more efficient in assimilating nutrients (Chisholm 1982; Seip & Reynolds 1995). The decrease of phytoplankton size at lower nutrient levels has been frequently observed in lakes (see, for example, Masson et al. 2000). The TP concentration dynamics of Lake Maggiore was accompanied by an increase in the relative biomass of small *r*-strategist phytoplankton (Bürgi & Stadelmann 2002).

The increasing importance of small r-strategists during an oligotrophication process has also been observed in Lake Michigan (Lehman 1991). Small algal taxa are expected to be favored under lower nutrient concentration, because their high SV-ratio enables rapid nutrient exchange through the cell surface (Harris 1994). The reduced grazing pressure could also explain the increasing number of small species. The top-down control (due to large herbivorous crustaceans) is important in deep lakes undergoing oligotrophication (Lehman 1991; Bürgi & Stadelmann 2002; Manca & Ruggiu 1998).

The Lake Maggiore TTSS pattern resembles the typical pattern established at Kinneret, a warm, meso-eutrophic lake. The two lakes are so different, but the stability of their phytoplankton taxonomic size structure is characterized by a comparable level of the similarity index (Pearson r here). A close level of similarity was also demonstrated by oligotrophic Lake Tahoe, USA (Kamenir et al. 2008b). In all three lakes (LK, LM, and LT), we can see 'more-or-less' the same size distribution resembling several bells symmetrical in logV scale. The TTSS fine structure can be described as several peaks separated by clearly seen troughs. While the peaks take almost the same horizontal positions (logV) across a time series, their height fluctuates from year to year (Figures 2 and 4).

If very close general patterns can be observed across long time intervals in very distant regions, one should look for very general mechanisms producing and supporting such distribution patterns (Havlicek & Carpenter 2001; Kamenir et al. 2008b). Specific studies are needed to select dominant factors and produce a pattern difference sufficiently big and consistent to be estimated quantitatively and compared with the change in the impact factors involved. We have considered one factor (very important, common and sometimes suitable to anthropogenic control), specifically, the nutrients level. The result gives evidence of the very high consistency of the TTSS general shape, accompanied by statistically significant differences in some peak heights, suitable to differentiation between two subsets of this pattern.

Lake Tahoe. Our study of LT phytoplankton also demonstrated high consistency of the TTSS pattern, found previously for LK. The similarity levels (Pearson r) for annual TTSS pairs for the Tahoe early period (1983-1986) are from 0.93 to 0.99, and for the following years (2002-2005) from 0.92-0.96. Such similarity levels are very close to r values of the stable period (1983-1986) of LK. While changes in the phytoplankton taxonomic composition took place at Lake Tahoe over that time period, and the species richness (number of OTUj) has dropped by more than 30% (Hunter et al. 1990), almost the same TTSS general pattern was evident (Figure 4). The similarity level between years 1983 and 2005 (i.e., time interval > 20 years) was still high ($r = 0.84$), comparable with the inter-lake (LK and LT) comparisons.

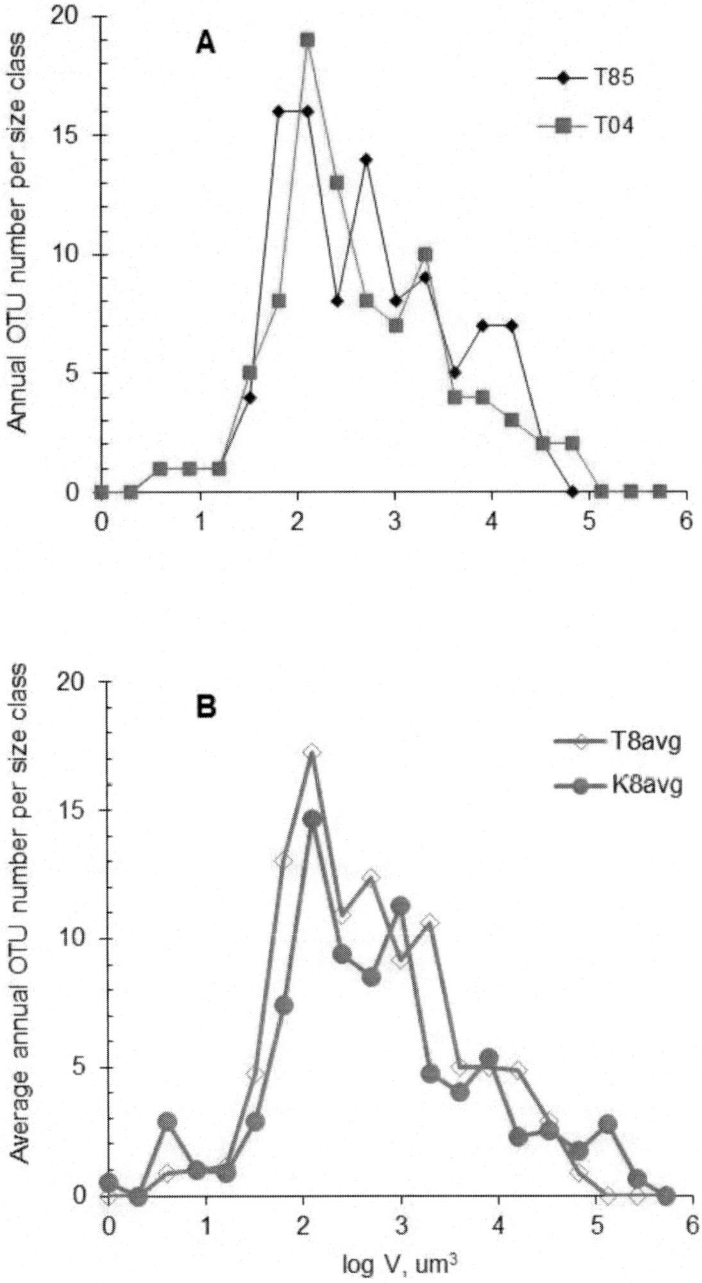

Figure 11-4. A comparison of traditional taxonomic size spectra (TTSS) of two lakes. A) TTSS-comparison of two years (1985 and 2004) of Lake Tahoe. B) Comparison of multi-annual average estimates (eight annual spectra obtained during two four-year periods) of two lakes (Kinneret and Tahoe, K8avg and T8avg, respectively).

The similarity level decline over long-time intervals is a result of height variations and side shifts of some peaks (Kamenir et al. 2006). This effect is clearly seen for Lake Tahoe (Figure 4), especially for the second (from the left) peak. A detailed analysis of the TTSS suggests that this 'shift' really reflects a change in the relative heights of several spectrum points. Namely, the notorious logV2.7 peak still exists in 2002-2003, but its height has decreased by a factor of two. At the same time, logV3.3 peak, which can be noted for 1984-1985 spectra has safeguarded its height and became more prominent. The logV2.7 peak drop was caused by the disappearance of many (>20) species of cell volume ~ 200-500 μm^3, most of which are Chlorophyta and diatoms. Disappearance of numerous taxonomic components, which may be important for floristic analysis, had almost no effect on the biomass dynamics and metabolic fluxes of the community. Additional collaborative studies of florists and mathematicians can be helpful in understanding the importance of such fluctuations of rare structural components.

At the same time, the most prominent (i.e., the highest) peak in both ecosystems and for all years (logV2.1) is one of the most stable, specific details (Figures 2 and 4). The left limb of this peak is particularly stable. The opposite side (the largest species) is evidently different between the lakes, as Lake Tahoe has no species of V > 100,000 μm^3 comparable with the large thecate dinoflagellates of subtropical Lake Kinneret. While the largest species body mass may be one important ecosystem characteristic (Kamenir & Shteinman 1999), this property also deserves additional attention. The peak/gap analysis has received much attention in recent times (Holling 1992; Allen et al. 1999; Havlicek & Carpenter 2001), providing an important way of understanding the community structural adaptation to its hierarchically structured environment. The statistical approach, profiting from the huge monitoring stores underexploited by many other methods, namely TTSS, can be an effective tool in such studies. A quantitative similarity measure (Pearson r here) transforms textual descriptions, common in the community taxonomic structure studies, into quantitative indices needed for mathematical analysis of such stability descriptors as deformation, robustness, and resilience (Odum 1971). A high similarity (r = 0.91) was found between the average (eight year groups) TTSS of the two lakes (Figure 4). This similarity is impressive in comparison with the annual TTSS variability (Figure 4). A possible explanation is that some self-maintenance mechanisms, which act within an aquatic community, tend to safeguard an optimal, and close to the symmetric lognormal bell, size-frequency distribution. Strong impacts cause deformations of the 'ideal' shape at all time-scales. The longer the study period the closer is the average pattern to its optimal shape.

Such reasoning speaks in favor of long-term studies carried out in more ecosystems. Main preliminary results (based on unpublished data) show that the general pattern resembles a Gaussian bell symmetrical in log (V) scale; there are, as a rule, 2-3 peaks in the bell central part, separated by more or less evident troughs. There are low secondary peaks in the left and right peripheral zones. The height of each peak and the total number of species vary from year to year. Such changes are markedly strong during long-term monitoring of oligotrophication and eutrophication processes. In shallow lakes benthic algae are registered in plankton samples, producing additional peaks in the far right region.

None of these details discredit the main conclusion of our study, i.e., the existence of consistent patterns of TTSS. While a large-scale comparative analysis is necessary, a 'contrasting pair' of lakes seems to be a reasonable interim stage necessary to prepare the grounds and develop a research program for a broad, multi-national comparative study. The TTSS pattern similarity (Figure 4A) is especially impressive in light of the fact that the laboratories worked independently over several decades. That is even more remarkable since various specific details of the data collection and processing were different and could have influenced the results.

The above results suggest that the long-term consistency of the aquatic assemblage taxonomic size structure pattern may be a common phenomenon. Its general pattern deserves more attention as a factor producing specific boundary limits, i.e., governing taxonomic composition adaptations carried on by self-maintaining communities. If established more precisely and confirmed by large-scale studies, it can become a valuable information means for both theoretical ecology and water-body monitoring applications. Taxonomic size spectra from terrestrial ecosystems (Holling 1992; Allen et al. 1999; Smith et al. 2004) suggest that common patterns may be established also for aquatic systems.

Our results demonstrate that the multimodal peak/trough structure is stable, withstanding pronounced changes in nutrient status, climate and even some artificial food web regulations. Close results were already described for experimental studies where lumps of taxonomic size structure had survived additions of nutrients sufficiently large to cause substantial changes of the plankton biomass and production. Lump regions were less self-similar if food web manipulations were added to the nutrient enrichments (Havlichek & Carpenter 2001). Similar size distributions were observed at several north temperate lakes with considerably different area, nutrient status, and trophic structure. The authors of the cited study concluded that taxonomic size distribution patterns may be conservative properties shaped not by lake-specific factors but by common regional ecosystem processes and organism patterns. Our results demonstrate that such patterns have similar properties even while observed during long time periods and in distant regions. Several mechanisms can be involved to explain the stability of the size distribution pattern. The multiplicative interaction of numerous impacts trying to destroy the structure of the optimal in some sense and feed-back supported system can lead to the lognormal general pattern (Chislenko 1981; Kamenir 1986). In a hierarchical structure – biosphere/ecosystems (Kamenir 1986) or taxonomic hierarchy (Chislenko 1981) - such process can produce a hierarchy of lognormal bells. The textural discontinuities in habitat structure can produce numerous modes (Holling 1992; Havlichek & Carpenter 2001) in this structure. The multi-lineage taxonomic background pool of our biosphere can be another way to support the observed multi-modality (Cumming & Havlichek 2002). Each of the above mechanisms can be thought as a member of a group of inter-complementary explanations. The most simple and general mechanisms can be the most appropriate to explain the long-time resilience of similar patterns in distant regions. A search for pronounced differences in the peak heights and horizontal positions holds potential to better understand the shaping mechanisms. Experimental studies, capable of causing the peak side shifts, are another efficient way to distinguish between the theories and to select operational models applicable to monitoring and ecological forecasting.

Plymouth marine station. The central point of our last comparison (a fresh-water lake and a marine ecosystem) is the phenomenon of high similarity in structural patterns shown by the phytoplankton assemblage taken as a whole. This similarity was estimated visually (graph comparisons) and with the help of formal quantitative methods (Pearson r). The similarity level obtained for marine phytoplankton was comparable to the levels found previously for lacustrine ecosystems. It is worth noting that such an approach helps us proceed from the frequently-used dichotomy ‘similar/dissimilar’ to a quantitative scale providing more opportunities for mathematical analyses of the obtained results. Such an approach helps us complement the first-level visual impression (the general pattern is practically the same for all studied years) with a more comprehensive judgment, taking into account numerous small modifications of specific size classes (Kamenir et al. 2010).

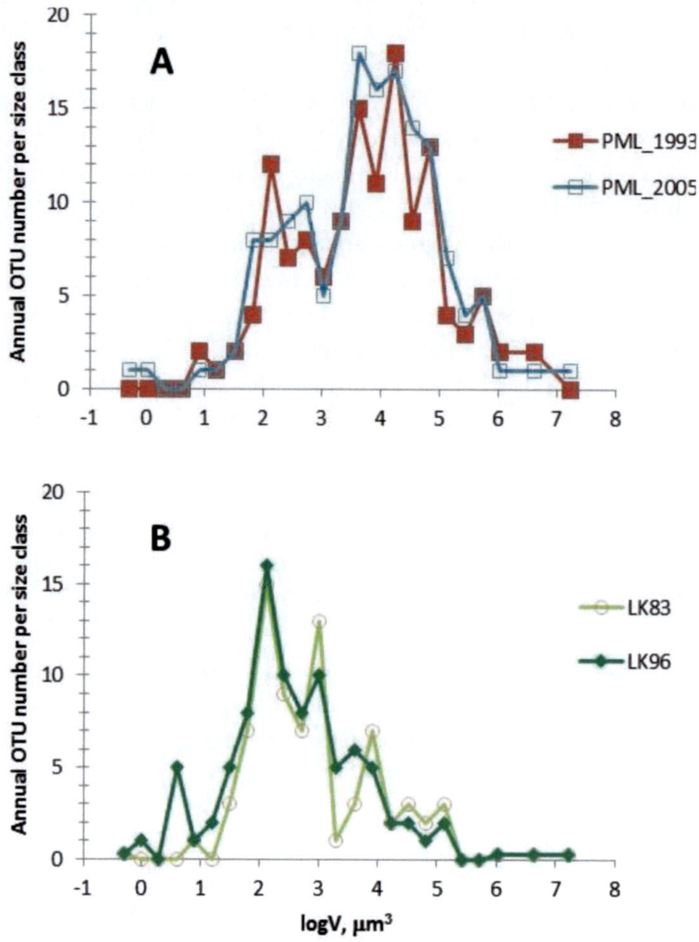

Figure11-5. Comparison of phytoplankton TTSS. A) The Plymouth Marine station L4 (PML) phytoplankton, two years (1993 and 2005). B) Lake Kinneret (LK), two years (1983 and 1996, typical and extremely disturbed years, respectively).

The much lower similarity level obtained in the inter-ecosystem pairs (LK and PML) is caused by very different positions of the second main peak (logV = 4.2 for the marine phytoplankton and 3.0 for LK). The peak side shifts seem to be quite probable during an ecological succession (Kamenir et al. 2006). Peak shifts were not noted for the L4 spectra, though numerous changes in the plankton of the western English Channel were documented during its long-term monitoring. The TTSS variability caused by such changes is especially well seen when comparing spectra divided by large time-intervals (Figure 5A). These changes demand special attention. Interdisciplinary analysis is necessary, as numerous environmental factors can be involved, including those pronounced over the last 100 years: climatic changes and anthropogenic disturbances (Hawkins et al. 2003).

If species size distributions prove to be a conservative property of terrestrial (Holling 1992; Allen et al. 1999) as well as aquatic communities (Warwick 1984; Roy et al. 2000), it demands new unifying principles in ecology (Havlicek & Carpenter 2001). The existence of ubiquitous species size-frequency distributions has already led to very fruitful theoretical discussions (Roy et al. 2000; Kozlowski 2002). Due to growing anthropogenic impact and global climatic change, environmental stability has become one of the most critical issues. While the term 'stability' has numerous definitions and properties, all of them depend on the system's self-similarity description and its dissimilarity quantitative estimation (Begon et al. 1996). Therefore, methods capable of describing biotic community structure, encompassing the whole community and quantitatively estimating its structural deformation, can be very important for the development of numerous fields in both theoretical and applied ecology. Our study describes just such a tool.

The phytoplankton TTSS pattern of a marine ecosystem (station L4) resembled the phytoplankton TTSS established for lakes (Figures 2, 4, 5B). The TTSS fine structure comparisons revealed numerous specific differences. Exampli gratia, the total phytoplankton size range was larger for the marine phytoplankton, where the largest cell volume was some 100 times higher than in LK. In both marine and lacustrine spectra, several secondary peaks can be seen. Our results demonstrate that the multimodal peak/trough structure of each ecosystem's annual TTSS is very rigid, despite pronounced changes in many environmental factors. For lakes, humps of taxonomic size structure survived the addition of nutrients sufficiently large to cause substantial changes in plankton biomass and production (Havlicek & Carpenter 2001). Taxonomic size distribution patterns can be conservative properties shaped not by lake-specific factors but by common regional ecosystem processes and organism patterns. Such patterns have very similar properties even for long time periods and in very different types of ecosystems.

Conservative multimodal distributions of this type have been studied for a long time in benthos (Schwinghamer 1981; Warwick 1984). The horizontal positions of the troughs and modes are among the most widely discussed topics (Warwick 1984; Brown 1995; Kozlowski 2002). Empirical data from numerous benthic communities, describing broad size ranges of organisms, show both the symmetrical (in logV scale) bells and the apparent, deep troughs dividing them. Such size spectrum properties also characterize marine phytoplankton, where similarity levels are comparable to those observed for phytoplankton of both oligotrophic and eutrophic lakes (Kamenir et al. 2008b). The similarity is strong even for time intervals of 10 years and more.

Numerous mechanisms characterized by sufficiently different time scales (ecological, micro- and macro-evolutionary), have already been suggested and analyzed to explain conservative properties of taxonomic size distributions. A homeostatic system would be characterized by a broad range of time delays and organism sizes (Kamenir 1986). A widely discussed formal mathematical model based on the trade-off between several metabolic processes (Brown 1995) was suggested for animal communities. While a realistic modal body weight estimate and the general distribution pattern discussed earlier (Hutchinson & MacArthur 1959) were obtained, the model (called BMT, based on the authors' names) also opened a way to numerous quantitative comparisons and formal discussions. Numerous problems, concerned with parameter interpretation and application at different levels of biological organization, were formulated (Kozlowski 2002). As the authors note, their model has not incorporated any inter-specific interactions or environmental relationships that may not be mediated by energetics (Brown 1995). Nevertheless, numerous consequences obtained via optimal body size analyses agree well with the empirical evidence.

Especially interesting for us are body size comparisons done for species from small islands and their nearest mainland relatives, and the decline in the species body size interval, which parallels landmass (continents and islands) area regression. Therefore, such regularities can be involved in explaining the above comparisons of maximal cell volumes and cell volume ranges (Figure 5A and B) done for marine ecosystems, with a small lake (e.g., LK).

Natural communities often include very different organism types, which can have different modal body weights. These differences are clearly seen in terrestrial mammals, birds, and fishes (Brown 1995; Smith et al. 2004). A community, where numerous high-level taxonomic units coexist, can have multimodal species size distribution. The relative height of each peak, and even its horizontal position (logV), can be changed by invasions or extinctions of species in each major taxon. While the spatial scale (ecosystem area) decline (from continental biota to large biomes and small patches of relatively uniform habitats) diminishes total biodiversity (Begon et al. 1996), high-level taxons respond by the modal peak height decline. The distributions become flatter and discontinuous. At the same time, similar body ranges are conserved, and an approximately equal number of species is observed in all non-empty logarithmic size classes (Brown 1995; Smith et al. 2004). If the total species number is small, some size classes become empty. Such effects are important for us as additional examples of multimodal curves. Water depth change can be very important if there are large shallow zones. Such a situation can be real for LK, where anthropogenic activities and several drought years have caused lake-depth and water-area reduction (Zohary 2004). Due to global climatic change, analogous situations can be expected during the following decades.

While numerous forces and mechanisms can affect the species size distribution of a given community, the final result depends on very general ones.

Each of the above explanations can be considered as a member of a group of interdependent and inter-complementary mechanisms. The most general mechanisms seem to be the most appropriate to consider while designing experiments. Since almost identical TTSS patterns were observed each year for English Channel phytoplankton taxonomic size structure, the TTSS multi-annual consistency can be helpful for ecological forecast. While TTSS takes into account very precious taxonomic information ignored by more popular biomass size spectra, it does not require taxonomic knowledge from the final user (a regional manager or environmental protection organization). As a set of reference values, the average TTSS pattern can be helpful in the everyday use of monitoring organizations, water body managers, and ecological modelers.

Conclusion

We conclude that the stability of phytoplankton taxonomic size structure documented for several aquatic ecosystems (LK, LM, LT, PML) is likely to have more general foundations. Results obtained from a lake (LM) oligotrophication study confirm our working hypotheses: reliable structural features were found in the TTSS general pattern of the Lake Maggiore phytoplankton, when many principal environmental characteristics changed considerably during the long-term oligotrophication process. The fine structure of TTSS reveals features helpful to divide the total oligotrophication process into several stages characterized with remarkable differences in the central peaks' proportions. Typical patterns of the stable and transitory stages were differentiated. Oligotrophication process decomposition was supported by quantitative statistical estimators.

The consistency of TTSS can produce quantitative operational monitoring tools acutely needed for ecological management, modeling and theoretical ecology. Such consistent patterns may be especially valuable as cornerstones needed for ecological estimations during times of environmental change. The TTSS fine structure variability can also lead to quantitative estimates for the typical pattern deformation. Such similarity/dissimilarity measures are necessary to produce quantitative interpretations for the aquatic assemblage stability studies. Since phytoplankton size structure provides much information about important physiological and temporal characteristics of the prime producer assemblage, the use of taxonomic size spectra can improve our ecological forecasting and interpretation abilities for many ecological systems, especially, while surveyed by modern monitoring means.

A comparison of highly different aquatic ecosystems (including a warm fresh-water lake and a marine phytoplankton assemblage) suggests that the long-term consistency of the aquatic assemblage taxonomic size structure pattern is a phenomenon leading to a valuable generalization. Such typical patterns deserve special attention during periods of accelerated global climate change and anthropogenic activity. TTSS multi-annual consistency can be helpful for ecological monitoring, modeling, and forecast. The evident and permanent differences in the peak positions between ecosystems open a way to additional analyses which can be helpful for the development of theoretical ecology and ecological forecast models.

The descriptors providing quantitative similarity estimations for aquatic assemblage structure (and not requiring deep taxonomic knowledge from the final user!) may be valuable for lake managers and environmental monitoring specialists. TTSS can be applied as a means of quantitative analysis for integral natural community structural evolution. Such an approach would facilitate application of the phytoplankton assemblage taxonomic structure in development of analytical tools acutely needed by environmental management, monitoring, and theoretical ecology.

Chapter 12

CONSISTENCY OF TAXONOMIC SIZE-STRUCTURE OF MAJOR PHYTOPLANKTON PHYLA

INTRODUCTION

A very important property of natural aquatic communities is their high taxonomic complexity reinforced by their spatial heterogeneity and seasonal dynamics. Analysis of such systems demands a huge amount of very competent work, which is often unrealistic. A rational solution can be found in a search for rigid estimates and patterns measurable by automated tools. If such rigid descriptors exist, they can be used as specific keystone reference points, to quantify the typical pattern deformations in the perturbed ecosystem (Kamenir 2011a). Hence, the ability to examine structural change in a living community or assemblage taken as a whole is highly important. Size spectrum (SS) analysis (Sheldon et al. 1972) is one of the tools capable to provide such comparative analyses. While the discussed species size-frequency distributions use the same structure as the biomass spectra (i.e., cell-volume logarithm axes and size-classes corresponding to the cell volume duplication), we prefer a more specific term – 'taxonomic size spectrum' (TSS). We have called it 'traditional' TSS (TTSS; Kamenir et al. 2006) to acknowledge its long history and to distinguish it from other modifications. One of such modifications - applied here - is the frequency-weighted taxonomic size spectrum (FTSS) (Kamenir 2007; Kamenir et al. 2008a).

We have already applied BSS, NBS, TTSS, and FTSS to analyze the phytoplankton of subtropical Lake Kinneret, Israel (Kamenir 2007; Kamenir et al. 2006, 2008b). The TTSS-shaping mechanisms have been studied for a long time, mainly for communities and taxons of terrestrial animals. Such studies have produced a set of models (Brown 1995; Kozlowski 2002) that do not contradict each other, but rather can be considered as intertwined, acting simultaneously and shaping the whole community (Kamenir et al. 2008a; Kamenir 2011). Taxonomic diversity of a natural community, producing a consistent TSS, is rather high – several dozens or hundreds species (Smith et al. 2004; Kamenir et al. 2008a). Hence, it seems logical that a population (i.e., a species or OTU) is just a small and replaceable element subjected to numerous kinds of stochastic impacts produced by the surrounding environment. Then a phylum, being an intermediate level between the whole community and its population, can demonstrate an intermediate consistency level. Hence, we assume that a simultaneous analysis of the whole assemblage and its parts can disclose a structural similarity decline across the hierarchical levels of a natural assemblage. While the high consistency in taxonomic and ataxonomic spectra of a community contradicts well-known facts of very high variability of the community taxonomic composition (Begon et al. 1996; Scheffer et al. 2003), simultaneous analyses of several hierarchical levels, including all community or assemblage and its major phyla, seem to be necessary. The TSS, capable of analyzing different levels of taxonomic organization and representing them as spectra, can serve as an efficient tool for such comparative studies (Kamenir et al. 2006). While the traditional TSSs produce rather crude assemblage descriptions for lakes and other small ecosystems, the frequency-weighted spectra can be more efficient in such studies. After C. Raunkiaer and H.A. Gleason, the number of sites occupied by the species is among the most common taxonomic descriptors (McGeoch & Gaston 2002). Nevertheless, no size-spectrum comparisons of this kind have been found in all the scientific literature known to us. Hence, we assume that the frequency-weighted spectra can bring highly novel and important results.

In this chapter, we describe and compare the size distribution patterns of the main phytoplankton phyla of a lake, as seen from long-time monitoring with the help of TSS amplified by the frequency-rate weighing of the assemblage taxonomic units.

The general aim of the study was a quantitative similarity assessment of the water mass occupancy annual patterns describing the size structure of Lake Kinneret (Israel) phytoplankton phyla.

The working hypothesis was that some consistent patterns of water-body occupancy by taxonomic components of phytoplankton will be obtained. The year-to-year similarity level of the whole assemblage structure was expected to be higher than such similarity level of its separate phytoplankton phyla.

MATERIALS AND METHODS

Data acquisition and processing were described above (chapters 8-10). The development of taxonomic size spectra was changed to compare two types of spectra traditional and frequency-weighted (TTSS and FTSS). A mean biovolume of the individual cell of each species, or size category for species counted under several size categories, was calculated from linear microscope measurements and the closest geometrical shape (Hillebrand et al. 1999). This cell volume was the parameter used for allocating a taxonomic unit to a size class. Since our individual taxon is not strictly a species but in some cases a size category within a species, we refer to each as an operational taxonomic unit (Sneath & Sokal 1973). Size classes were created by the cell size logarithm increments of 0.301. The Vxx notation is used throughout this paper for size classes, where xx is the logarithm of the class' right boundary (Kamenir et al. 2008b).

Each TTSS was built as the discontinuous frequency distribution (histogram) of the total number of OTUs to size classes (Figure 1A). Eight annual TTSSs, one for each year in the two 4-year sets, were built and analyzed previously (Kamenir et al. 2006) using the same method. The frequency-weighted taxonomic size spectrum (FTSS) has been invented and was simultaneously applied to study the same 8 years (Kamenir et al. 2008a).

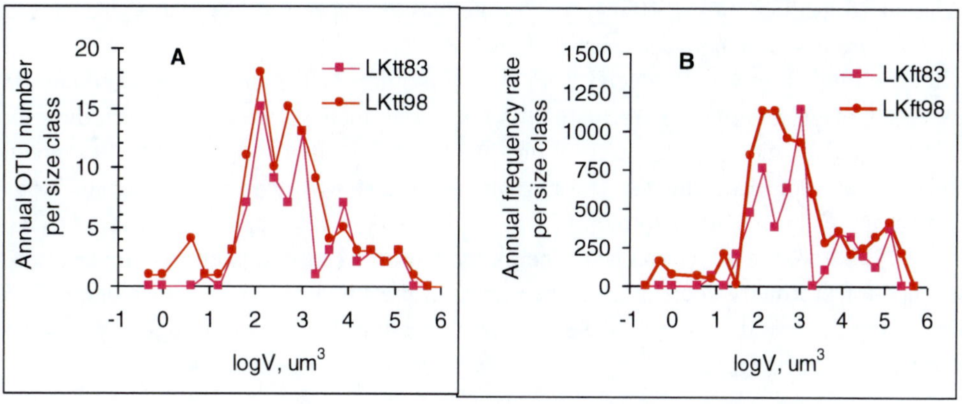

Figure 12-1. Comparison of annual taxonomic size spectra. A, B) The traditional (TTSS) and frequency-weighted (FTSS) spectrum change, respectively. Representative years of the stable and extremely disturbed periods (1983, 1998, respectively; after Kamenir et al. 2008a). While the traditional spectrum (LKttXX, where XX is the year number) shows the distribution of operational taxonomic unit (OTU) number between the cell volume logarithmic size classes, the modified spectrum (LKftXX) applies to each OTUj its frequency weight (FRj; see the Methods section). Annual FRj is the total number of water samples where respective OTUj was registered during the studied year. Size classes were created by doubling cell volume (V), and are presented on a logarithmic scale ($\log_{10}V$).

While TTSS ignores the time-spatial structure of the assemblage, the FTSS applies to each OTU a 'weight' – its frequency rate (FRj) – which is the number of water samples where such OTU was registered. Not cell abundance is considered here but only OTU incidence, i.e., whether an OTU occurred in the annual database (TTSS) or in a specific sample (FTSS). Frequency rate values were plotted along the Y axis of the FTSS histogram (Figure 1B). These values are the total numbers of times all OTUs which belong to a specific size class (*i*) were registered during the year. The 'team score' of each size class (FRRi) integrates the FRj-weights of all OTUs falling into size class *i* (FRji), according to:

$$FRRi = \Sigma\ FRji$$ (Kamenir et al. 2008a).

Using the histogram and crosstab procedures, we calculated FTSS as the size-frequency distribution of all database records. In the present study, analogous spectra (FTSS) were prepared for other years, from 1986 to 1995. The SSyy notation was applied to refer to the SS of a particular year (yy), for example, FTSS92 was the FTSS for 1992. A pair of annual spectra (TTSS and FTSS) were developed in the current study, each year, for each of eight major taxonomic groups (phyla) of Lake Kinneret phytoplankton, which are Cyanophyta, Diatomea, Chlorophyta, Dinoflagellata, Cryptophyta, Prasynophyta, Euglenophyta, and Haptophyta (cya2, dia3, chl4, din5, cry6, pras7, eug8, and hap9). The 15-year interval (1985 - 1999) was divided into three 5-year periods. Then, for each of those 5-year periods, average values for each size class were calculated from the FTSS sets (Kamenir et al. 2008a).

Similarity estimation - Treating a histogram (TTSS, or FTSS; a discontinuous distribution function) as a numerical vector, we can estimate the similarity/dissimilarity measure for each pair of spectra, using hierarchical cluster analysis (Cha & Srihari 2002). The Pearson correlation (Sneath & Sokal 1973) was applied as the spectrum similarity measure. The estimates were based on between-group linkage.

A separate dataset was produced from each proximity matrix describing a phylum or the whole assemblage. Fourteen elements of the respective matrix 1st column, excluding its 1st trivial diagonal element, were used. While parametric method applicability has not been established for such data, the Wilcoxon signed-rank test was applied to compare a pair of such datasets. Fisher's Z-transformation was applied to the correlation coefficients (Sokal & Rohlf 1995) before the Wilcoxon test. The paired t-test was also applied to compare two sets of seven Z-transformed *r* coefficients, where each of the studied years was included only once (years 1 and 2, 3 and 4, etc.). SPSS program, version 15.0, was used for all statistical procedures.

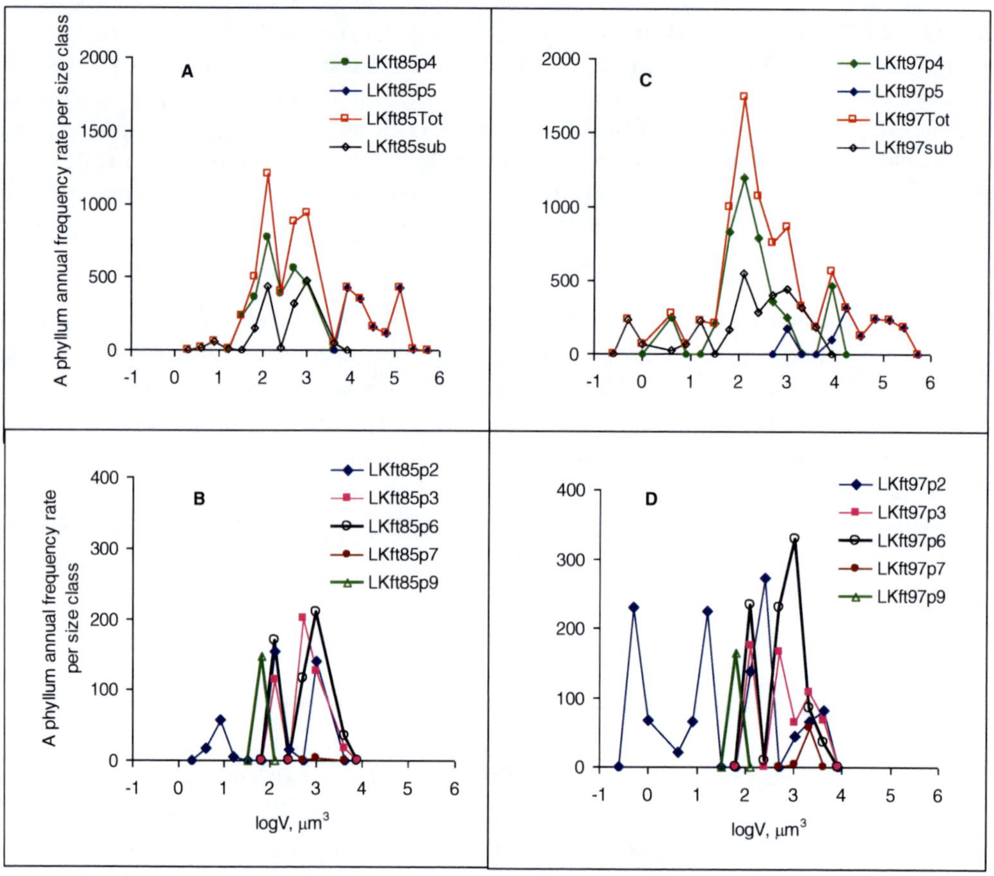

Figure 12-2. FTSS comparison for the major taxonomic groups (phyla) of Lake Kinneret phytoplankton: Cyanophyta, Diatomea, Chlorophyta, Dinoflagellata, Cryptophyta, Prasynophyta, Euglenophyta, and Haptophyta (cya2, dia3, chl4, din5, cry6, pras7, eug8, and hap9, respectively). LKftXXpY means Lake Kinneret FTSS of year XX (1985 and 1997) for phylum Y. "Tot" describes the whole phytoplankton assemblage, while "sub" stands for the sum of all subdominant phyla represented in the respective lower plate. Some phyla exhibit pronounced differences between the two years (see details in Table 1). Note the Y-axes scale difference between the upper and lower plates.

Table 1. Quantitative comparison of FTSS for seven major taxonomic groups (phyla) of Lake Kinneret phytoplankton*

	Cyanophyta - cya2		Diatomea - dia3		Chlorophyta - chl4		Dinophyta - din5		Cryptophyta - cry6		Prasynophyta - pra		Haptophyta - hap9		Total
LogV	1985	1997	1985	1997	1985	1997	1985	1997	1985	1997	1985	1997	1985	1997	Frequency
-0.3	0	**230**	0	0	0	0	0	0	0	0	0	0	0	0	230
0.0	0	68	0	0	0	0	0	0	0	0	0	0	0	0	68
0.3	0	0	0	0	0	0	0	0	0	0	0	0	0	0	0
0.6	18	21	0	0	0	**249**	0	0	0	0	0	0	0	0	288
0.9	58	65	0	0	0	0	0	0	0	0	0	0	0	0	123
1.2	6	**225**	0	0	0	0	0	0	0	0	0	0	0	0	231
1.5	0	0	0	0	235	210	0	0	0	0	0	0	0	0	445
1.8	0	0	0	0	356	**828**	0	0	0	0	0	0	147	164	1495
2.1	154	138	115	176	767	**1195**	0	0	169	234	0	0	0	0	2948
2.4	15	**272**	0	0	384	**792**	0	0	0	9	0	0	0	0	1472
2.7	0	0	201	167	559	358	0	0	116	**230**	0	0	0	0	1631
3.0	**141**	45	**126**	64	**458**	246	0	**176**	210	330	4	4	0	0	1804
3.3	0	**66**	0	**108**	0	5	0	1	0	**85**	0	**57**	0	0	322
3.6	0	**82**	17	67	0	0	0	0	35	35	0	0	0	0	236
3.9	0	0	0	0	0	**468**	**425**	97	0	0	0	0	0	0	990
4.2	0	0	0	0	0	0	349	317	0	0	0	0	0	0	666
4.5	0	0	0	0	0	0	162	127	0	0	0	0	0	0	289
4.8	0	0	0	0	0	0	113	**244**	0	0	0	0	0	0	357
5.1	0	0	0	0	0	0	**426**	229	0	0	0	0	0	0	655
5.4	0	0	0	0	0	0	11	**186**	0	0	0	0	0	0	197
Total	392	**1212**	459	582	2759	**4351**	1486	1377	530	**923**	4	**61**	147	164	14447

*The groups are Cyanophyta, Diatomea, Chlorophyta, Dinoflagellata, Cryptophyta, Prasynophyta and Haptophyta (cya2, dia3, chl4, din5, cry6, pras7 and hap9, respectively). Some phyla have pronounced differences (marked bold) between the two years (see Figures 2 to 5). The border delimits 'the central size range' where almost all groups are represented. For each size class, the frequency rate represents the sum of all OTUs belonging to it. The frequency rate of each OTU represents the number of samples where it was found during the respective annual period. The frequency rate growth of all phyla in 1997 is concentrated in almost non-overlapping zones: cya2 has changed 3 times, mostly at V0, 1.2, 2.4, 3.3-3.6; chl4 ~1.5 times, at V1.8-2.4 and 3.9; cry6 ~ two times, at V2.7-3.3; din5 has changed in its most peripheral zones of V3.0 and in highly important maximal cell-volume zone (V5.4) where its frequency has grown 17 times! Euglenophyta (eug8, several samples registered for Kinneret OTUs of logV = 2.7-4.2) were not registered during 1985 and 1997.

RESULTS

The assemblage and its main phyla - The FTSS curves for two of the 15 years, specifically 1985 and 1997, are shown in Figure 2A-D, comparing all phytoplankton phyla found in the Kinneret. These two years represent the longtime 'typical' and the more recent disturbed periods, respectively. For each comparison, we distinguish two stages of the spectrum description: 1) its general pattern ignoring slight variations of specific peak heights and side shifts; 2) the spectrum fine structure quantitatively describing each peak and trough. Both FTSSs of the whole assemblage (index 'Tot' at the curve-symbol end) are similar in their general shape (Figure 2A and C), showing peaks and troughs in the same size classes.

The 1997 spectrum is some 40% higher, with a stronger domination of the first (V2.1) peak, representing cells of some 100 um^3. This spectrum also demonstrates the addition of a pair of low peaks for cells < 10 um^3, and a lower but broader zone describing the largest algae (logV > 4). The specific phylum spectra help us better evaluate these changes.

The two dominant (according with FRj criterion) phyla are chl4 and din5 (Figure 2A and C). The two curves describing them show a very small overlap. While the second group (din5) has no companion in the size range logV > 4, its FTSS coincides there with the FTSS of the whole phytoplankton. In a strong contrast, the central region (logV = 2-3) demonstrates evident differences between the dominant group (chl4) and the whole assemblage curve. The reason for this is that all other phyla (cya2, dia3, cry6, pras7, and hap9) are also represented inside the central region (Figure 2B and D).

Tabular comparisons (Table 1) help us better notice numerous small details in the FTSS fine structure of some phyla that would not be easily noticed in the crowded graphs of Figure 2. Two of the phyla (dia3 and cry6) have rather similar typical patterns of 2-3 peaks divided by deep troughs and encompassing a cell volume range of 1.5-2 orders. Two others (pras7 and hap9) have very low frequencies; each of them produces only one very thin peak. A notable exception is cya2 with 3 in 1985, and even 4 in 1997, peaks of almost the same height, encompassing a much broader size range (logV = -0.3 to 3.6). Din5 also has four peaks (though only two of them are high) and a broad size range from V3.0 to V5.4. One can see that the total activity (last line, Table 1) has grown >3 times for cya2 (1997 vs. 1985). For chl4, the difference was less pronounced – some 40%, for cry6 – some 80%, for dia3 and hap9 – it was negligible, and for din5 – it was negative and small. The greatest change (>15 times) was obtained for pras7, but its absolute values were very low, being produced by just one species.

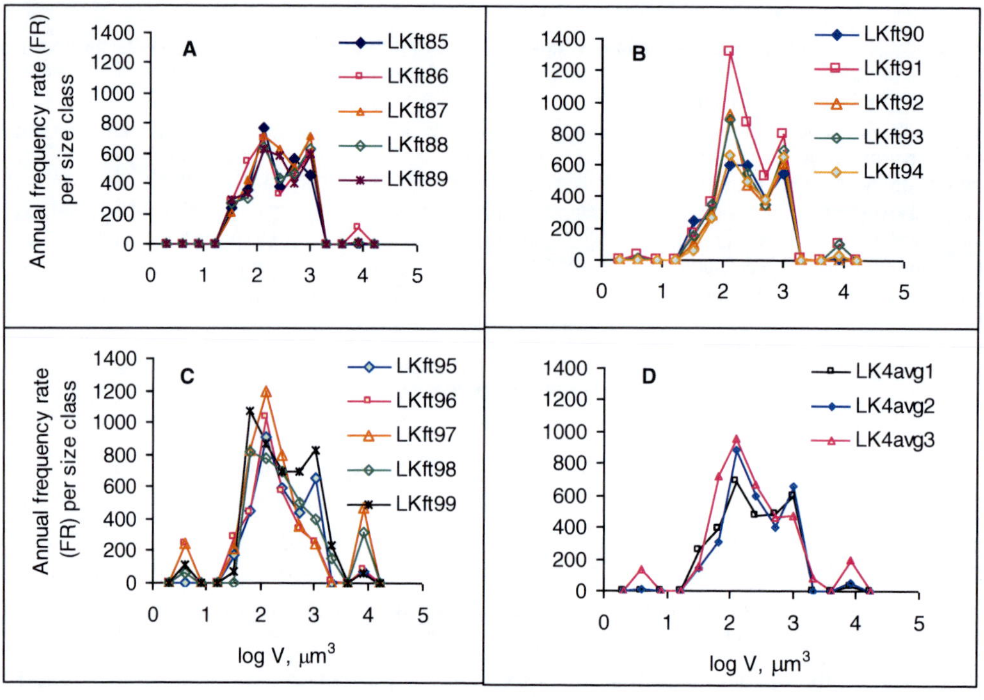

Figure 12-3. Comparison of FTSS annual patterns for one phytoplankton phylum (Chlorophyta, chl4) of Lake Kinneret; 3 consecutive periods of 5 years each. Two main and two smaller peaks are distinguishable. Changes of the peak height, breadth, and even small side shifts are evident. The symbols are described in the legend of Figure 2. LKxAvgY means the average spectrum of a 5-year period Y (Y = 1 to 3) for phylum "x" (see Figure 2); LK4 (i.e., x = 4) means chl4.

Closer attention highlights many variable details in the spectra of specific phyla (Table 1). The pronounced change of cya2 FTSS was produced mostly by the growth of several previously unimportant size classes (specifically, V-0.3, V1.2, and V2.4, producing high peaks) and the expansion of the V3.0 peak rightwards, up to V3.6. The increase in the total frequency rate transformed this group (cya2) into a sub-dominant, comparable to the second dominant (din5). In the same way, the dominant phylum (chl4) added two size classes (separate side-peaks) at V0.6 and V3.9 (Table 1). It also played an important role in the central region deformation via upgrading its main peak (V2.1) while diminishing the second peak (V3.0). Din5 - the biomass dominant of the Kinneret phytoplankton (Zohary 2004) - also had pronounced changes in its FTSS pattern (Table 1). Its peak classes, V3.9 and V5.1, decreased 2-4 times, while its right extremity V5.4 has grown some 17 times!

Three consecutive periods of the main phyla - Pattern variability was more noticeable when comparing a set of spectra (15 consecutive years) for each phylum. The most dynamic phylum was cya2 (to be described in a special publication), though changes of the same type can be seen in each of phyla (Figures 3 to 5). One can see that the three consecutive 5-year intervals demonstrate different levels of variability: (Figures 3 and 4), the first period (1985-1989) demonstrated only the peak height fluctuations, while the peak horizontal positions and shapes were the same for all years (Figure 3A). During the second period (1990-1994), some peaks changed their shapes (Figure 4B).

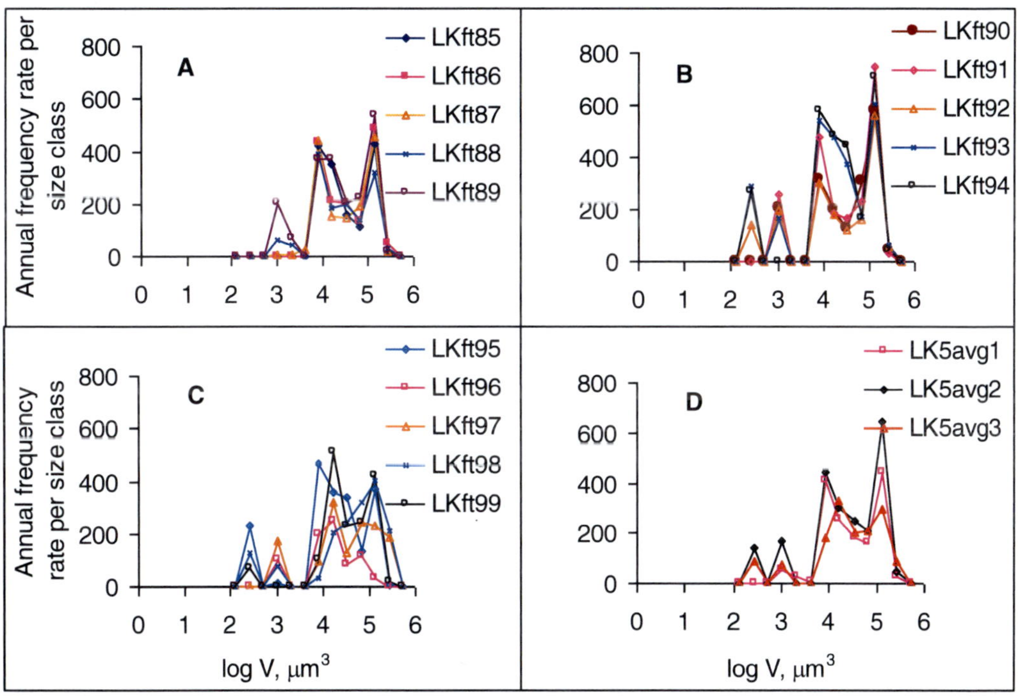

Figure 12-4. Comparison of FTSS annual patterns for one phytoplankton phylum (Dinophyta, din5) of Lake Kinneret; 3 consecutive periods of 5 years each. Two main and two smaller peaks are distinguishable. Changes of the peak height, breadth, and even side shifts are seen. All symbols are described in the legends of Figures 2 and 3.

The third period (1995-1999, Figures 3C and 4C) demonstrates a much higher variability, confirming the judgments produced by more traditional methods (Zohary 2004). The peak-height variability was growing, and the last years (1995-1999) even showed slight side shifts. A special comparison of the 5-year average spectra of these three periods makes these differences more obvious (Figure 3D).

The din5 comparison demonstrated very similar properties (Figure 4). Only weak fluctuations were noted during the first period (Figure 4A), but its last year (1989) demonstrated a significant height rise of a peripheral peak at V3.0. During the second period, much stronger changes were seen, mostly in the V4.2-V4.8 and V2.4 regions (Figure 4B). The third period (1996-1999) demonstrated even more pronounced FTSS deformation, including the change of the maximal cell-volume zone (Figure 4C-D).

A variability comparison - A simultaneous comparison of the 5-year average patterns of the five main (Figure 2; Table 1) phyla (cya2, dia3, chl4, din5, and cry6) demonstrates evident peak height variability (Figure 5). As a rule, the phylum peaks are separated by a deep trough and are divided by 1-2 orders of magnitude of cell volume. The horizontal positions of the main peaks are different for different phyla. Din5 and cya2 are especially interesting in this respect, taking advantage of the extreme size ranges unused by all other phytoplankton phyla (Figure 5). While the fine-structure variability is evident from the phylum FTSS dynamics, the 'traditional' spectrum (TTSS) seems to be less valuable for phylum analysis. The low gradation number in each size class is noted in TTSS (Figure 5A) during the stable years (till 1994) even for such a large phylum as cya2 (Figure 2; Table 1). The phylum FTSS provides more precise spectrum profiling than its TTSS (Figure 5B and A, respectively).

The dendrogram and its proximity matrix were produced for each of the five main phyla and the whole assemblage, all dendrogram trees providing very close clustering of the years. The lowest similarity (Pearson r) estimates reflected strong FTSS pattern deformations during the extreme years, such as apparent peak proportion changes and peak side shifts (Figures 3C and 4C). The difference in the FTSS similarity level between the most variable phylum (cya2) and the whole assemblage was statistically significant (Wilcoxon signed-rank test; $z = 3.30$, $n = 14$, $p = 0.001$). This significant difference was supported also by the paired t-test ($t = 8.47$, $df = 6$, $p < 0.001$).

DISCUSSION

Typical patterns - The central point of this study is the phenomenon of high and long-term consistency in structural patterns of the lake phytoplankton major parts (phyla) during a long-term period of time. Such analysis considers not only the whole assemblage but adds a specific comparison between major components of the next level of its hierarchy. Specifically, we see consistent patterns of the water body occupancy by size fractions and taxonomic groups. Each phylum (cya2, dia3, chl4, din5, cry6; Table 1) has its characteristic pattern (Figure 5) and is distinguishable from other groups of the same level of taxonomic hierarchy.

This FTSS property – consistent patterns – is already known for several other types of size spectra: BSS, NBS, and TTSS. In this study, consistent FTSS patterns were established for Lake Kinneret during a period of 15 years. Even for the extreme years (1996-1999), the general FTSS patterns resembled those established during the previous stable period. At the same time, the FTSS deformations were more explicit during the extreme years. The FTSS deformations were more pronounced for some phyla than for the whole assemblage. The most interesting one (in the Kinneret case) seems to be cya2 (Figure 5B). Its pronounced deformations during the last years of the studied period produce very low Pearson r estimates.

Therefore, the full r-range of cya2 is >5 times broader than the r-range of the whole phytoplankton. This property may be diagnostically valuable for the water-quality maintenance and ecological monitoring.

The difference between the assemblage and its phylum consistency level was statistically significant. We interpret this difference as evidence of a specific self-maintenance mechanism.

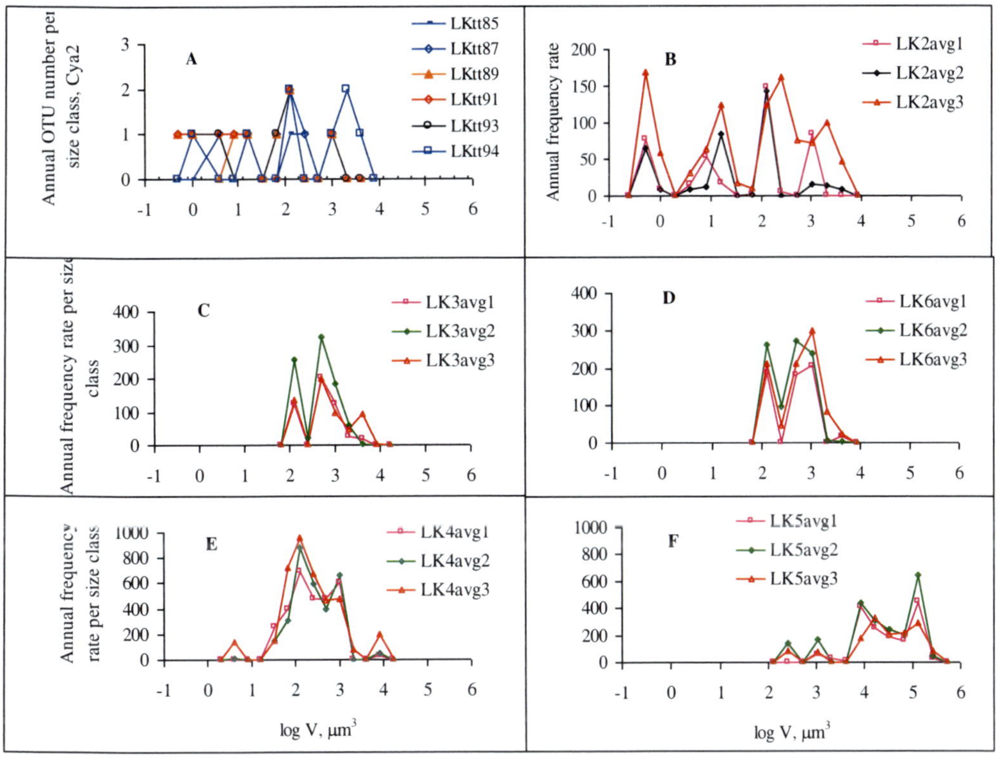

Figure 12-5. Comparison of FTSS annual patterns for five main phyla of Lake Kinneret phytoplankton; 3 consecutive periods of 5 years each. Plate A represents the TTSS variability of one phylum (Cyanophyta, cya2). The phylum TTSS is much more crude than the respective phylum FTSS (plate B), as the annual change is typically only 1 species addition/deletion per size class. All symbols are described in the legends of Figures 2 and 3.

It seems to be a community-level mechanism: while all ecosystem components function under stochastic environmental conditions, each population and phylum express incessant fluctuations. At the same time, the phyla appear to be 'tied together' by the higher-level (the whole phytoplankton) size distribution limits. These limits can be described as a lognormal or a hierarchy of lognormal distributions (Figure 1). Log-normal functions are already often applied to approximate species diversity distribution versus organism size in terrestrial and aquatic communities, including plankton (Irwin et al. 2006). The whole assemblage seems to be rather stable: its FTSS deformations become evident only during the extreme disturbance periods (Figures 1B, 5B). At the same time, even during the stable period, some species demonstrate almost chaotic dynamics unsuitable for making forecasts. Many of them even do not appear (are not registered) each year. Hence, the phylum appears to be an intermediate level of optimal sensitivity. Safeguarding its FTSS general pattern, the phylum demonstrates deformations of its spectrum fine structure, quite evident and suitable for quantitative estimations even under moderate environmental disturbances of the relatively stable period (Figures 3 and 4).

Such patterns of taxonomic size structure, consistent during long periods of time, help us settle a conflict between the existence of ataxonomic size spectrum (BSS and NBS) typical patterns (Platt & Denman 1978; Sprules & Munawar 1986, Kerr & Dickie 2001), and the widely known (e.g., Scheffer et al. 2003) high flexibility of the community taxonomic composition (i.e., the list of species producing these common and stable ataxonomic size distributions). The size-frequency distribution of species (TTSS) - that is already common in terrestrial zoology (Chislenko 1981; Smith et al. 2004) - is rare in hydrobiology (e.g., Roy et al. 2000; Havlicek & Carpenter 2001), and the FTSS was never found in the literature known to us. Nevertheless, both these 'rare tools' can be efficient and help us relate the 'ataxonomic' and the taxonomic size structures. Moreover, they can amplify our diagnostic capabilities, adding the fine-structure comparisons of some intermediate hierarchical levels, i.e., of specific taxonomic groups selected for the study.

The central region - The SS matrix, composed of several lines, each representing a specific size spectrum type, can be a more efficient community-structure descriptor than a separate index or one spectrum (Kamenir 2007). Such a matrix can make our description of the aquatic community structure more comprehensive and help us obtain a better understanding of the community adaptation mechanisms. While the NBS emphasizes the importance of small organisms (as the most abundant ones), the BSS is 'dominated' by large organisms (producing high biomasses).

Taxonomic spectra (TTSS, FTSS) give most attention to the central region noted for the highest taxonomic richness. In our case, the FTSS center contains >50% of all species, while the biomass-producing right extreme zone (logV>5) contains only 2 OTUs (Figure 1A).

Only half of the phytoplankton phyla commonly found in freshwater (John et al. 2005) maintain species numbers and size ranges producing more than one peak in Lake Kinneret TSS. All these phyla are represented inside the narrow central region (logV = 2.1-3.0; Figure 2; Table 1). Phyla with low diversity (pras7 and hap9) are located at the borders of this narrow region.

On the other hand, almost all phyla have their highest peaks at different locations. Only the most prominent central peak (V2.1) is 'preferred' by cya2, chl4, and cry6 (Figure 2; Table 1). Considering the cell volume – one of its most informative characteristics (Peters 1983; Irwin et al. 2006) - as an ecological-niche descriptor, one can assume that the main phyla tend to a 'niche-separation' minimizing the TSS peak overlaps (Figure 2; Table 1). Especially impressive is this comparison for din5 and chl4, i.e., the two phyla dominating the Kinneret phytoplankton biomass and its species number, respectively. For the four smaller groups (dia3, cry6, pras7, hap9), the conflict between the two trends (i.e., of getting closer to the distribution center while trying to minimize the peak overlap with all other groups) results in a crowded 'package' of their main peaks within a very narrow band of ~1 order of cell-volume change. The notorious cya2, profiting from the free size range of V < 1.5, exploits also the central region at V2.4, between the main peaks of dia3 and cry6 (Figure 2D).

Each of the three most minor taxa (pras7 and hap9; eug8 registered during only 1 year of the period described here, i.e., in 1999) is concentrated in just one size class. Excluding these three smallest groups, encompassing 4% of species and 1.6% of the total frequency (i.e., registration records), the 'central region' can be defined as 4 size classes (V2.1-V3.0; Table 1). This strip – only 20% of the total range - contains >50% of all species (some 200 taxonomic units observed during 15 years) and their registration records (Figures 1 and 2). Dinophyta seems to be very interesting in relation to this trend, as it is located mainly outside of the central region, while profiting from the free logV > 4 range producing the main part of the phytoplankton assemblage biomass.

The pattern-shaping mechanisms - Several mechanisms that can be involved in explaining the long-term consistency of the size-distribution pattern were considered by us previously (Kamenir et al. 2008b). The multiplicative interaction of numerous impacts can produce lognormal distributions of the feed-back supported characteristic optimized by some ecological mechanisms.

Such mechanisms are already discussed for taxonomic size-frequency distributions described in zoology (Brown 1995; Roy et al. 2000; Kozlowski 2002). In a hierarchical structure – biosphere/ecosystems or taxonomic hierarchy – such a process can produce a hierarchy of lognormal bells (Chislenko 1981). Then, the textural discontinuities in the habitat structure and multi-lineage taxonomic background pool can produce numerous modes (Havlicek & Carpenter 2001; Kamenir et al. 2008b). While all these mechanisms can act simultaneously, each of them can be thought of as a member of a group of inter-complementary explanations. The most simple and general mechanisms seem to be the most appropriate, as the same patterns are found during relatively long periods of time and in very distant regions, such as Lakes Kinneret (Israel) and Tahoe (California) (Kamenir et al. 2008b).

Very close results were obtained for much larger time and space scales. Published paleontological data from several oceanic provinces show that for marine bivalves latitudinal change patterns of species richness are decoupled from patterns of species size-frequency distributions (SFD), which can be considered a fundamental aspect of species ecology and life history (Roy et al. 2000). Regional SFDs are statistically indistinguishable, despite significant changes in species richness and in numerous variables that can influence body size, such as temperature, seasonality, and ecosystem productivity. SFDs of bivalves in the northeastern Pacific demonstrate remarkably stable patterns over 5,000 km of continental shelf, from the equator to the polar sea. The patterns safeguard their main parameters (body size range and its modal value) over periods of several million years. Both the mode and shape of these invariant SFDs are successfully predicted by the known BMT model based on energetic and life-history parameters (Roy et al. 2000). Though we analyze a considerably different type of organism occupying another geographical zone and trophic level, and we also analyze quite different scales of space and time, the above conclusions seem to be quite appropriate to describe our results. Therefore, the TSS long-term consistency seems to be a very general phenomenon of a very general nature, produced by mechanisms acting at different scales in time and space. Thus, the time-space scales of our science (i.e., a water body of several miles undergoing monitoring for several years) provide huge advantages in comparison to the above paleontological studies.

The multi-level structure - While analyzing the long-term consistency of the total assemblage, we can consider and compare FTSS at several levels of its hierarchical structure

(Figure 2). Such comparisons support the whole-object analysis based on the dynamics of its most prominent spectral zones (Figures 2 to 5) and then the dynamics of particular species or OTUs producing a certain peak. While all classes of all phyla fluctuate from year to year (Figures 3 to 5), the total number of OTUs and their registration records also fluctuate. Even so, the FTSS general pattern of the whole phytoplankton assemblage is almost the same (Figure 2), while its specific phyla show much greater pattern variability (Figures 3 to 5). The dynamics of separate species is even more chaotic and less predictable. In this respect, our long-term monitoring results resemble those of a long-term laboratory experiment (Heerkloss & Klinkenberg 1998) of a species-depleted plankton mesocosm. Especially dramatic changes were evident in our study for some size classes of Dinophyta and Cyanophyta after 1994 (Figures 3 and 4). The dynamics of each size spectrum peak of each phylum is produced in our case by a rather small number ($n < 10$) of species. Therefore, we can go down and analyze each of them. At that level we see a rather chaotic dynamic of some populations. We can note many 'sporadic' species registered only during short periods of time, to be replaced by other 'sporadic' ones. While such 'small-group chaos' strongly resembles laboratory experiments (Beninca et al. 2008), it is replaced by rather consistent quotas at the next level (a phylum size class here) and by almost steady patterns of the whole assemblage size spectrum (Figures 3 to 5). A pronounced decline in such structural similarity emerged only for some taxons and only during the years described (Zohary 2004) as extremely unusual (Figure 5). We interpret this coexistence of the orderliness and chaos (Scheffer et al. 2003) as evidence of a specific statistical mechanism being more efficient at higher levels of hierarchy, encompassing higher numbers of species. The simultaneous analysis of several hierarchic levels of the aquatic assemblage helps us find the most pronounced deformations of specific size ranges, which are produced by specific taxonomic components – phyla (Figures 3 to 5). In turn, those changes are produced by particular species. For example, the pronounced deformations of peak V2.1 were produced mainly by just two cya2 species - *Microcystis flos-aquae* and *Chroococus minutus*. FTSS changes of cya2 are very pronounced (Figure 2B and D; Table 1), conforming to its biomass changes described previously (Zohary 2004). If the community functional parameters adapt much later than the community taxonomic composition (Zohary 2004), FTSS may be applied as an early warning diagnostic tool. This tool helps us encompass the whole assemblage, compare its hierarchical levels, and find its components (phyla, size fractions, and species) most informative for diagnostics of community reorganization under analyzed disturbances. Being based on large datasets and OTU methodology, FTSS can be especially valuable for automated monitoring. Modern plankton counters, flow cytometers and image analyzers are already capable of automating almost all data collection and taxonomic classification with the speed of some 10,000 particles per hour (Sosik & Olson 2007). Therefore, such an approach opens prospects to obtain huge data volumes, up to the levels demanded by specific mathematical models and statistical theory. The sufficiently higher precision of the quantitative estimates, needed for particular taxons and mathematical models, becomes a reality.

All in all, we see a multi-level stabilization system, where very high flexibility of the taxonomic composition (species list) parallels the low variability of the functional parameters of the whole object (the ecosystem and its living community), e.g., its total biomass, annual primary production and community respiration (Scheffer et al. 2003; Zohary 2004). This property is realized through a community-level taxonomic size-distribution consistency linked to stable ataxonomic distributions supporting the community-level flows of energy and nutrients. The whole-community size-distribution patterns produce size-class limitations for numerous species attempting to find a dwelling place inside the ever-unsteady environment of our ecosystem.

CONCLUSION

Quantitative similarity comparisons demonstrate annual patterns of water mass occupancy by Lake Kinneret phytoplankton assemblage and its main phyla, consistent during a long-term (15 years) period.

The consistency of the FTSS *general pattern* was proven for all major phytoplankton phyla during the 'extreme' period (1996–1999), characterized by pronounced deformations of the annual succession pattern established during the previous 'stable' period (1985-1995). Simultaneously, specific changes in the FTSS *fine structure* and a strong decline in the FTSS year-to-year similarity level (Pearson r) were evident for all phyla and may be helpful for diagnostics, being more pronounced at the phylum level than for the whole phytoplankton assemblage.

Thus, quantitative similarity comparisons prove that the whole-assemblage consistency level is superior to that of its major parts (phytoplankton phyla here). This relation confirms our hypothesis on the whole-community origin of the statistical mechanisms producing long-term consistency of taxonomic size structure.

The FTSS profiling precision can be upgraded further with the help of automated analysis based on OTU methodology. The descriptors providing quantitative similarity comparisons for the aquatic assemblage structure, while not requiring a deep taxonomic knowledge from the final user, may be helpful for lake managers and environmental agencies.

Chapter 13

A MULTISIDED ANALYSIS OF LIVING MATTER

INTRODUCTION

Practical ecology often focuses attention on species identities. Such an approach becomes very complicated when real ecosystems encompassing dozens or hundreds of species are studied. The trait-based approaches (Litchman et al. 2007; McGill et al. 2006) try to overcome this obstacle by concentrating on formal quantitative properties of living organisms. An alternative to taxonomy-based schemes is the living matter (LIM) concept (Vernadsky 1978), which encompasses the whole assemblage of co-occurring living cells.

An ecosystem is a community of living organisms in conjunction with several abiotic factors of the environment, interacting as a system. These biotic and abiotic components are linked through nutrient cycles and energy flows (Odum 1971; Begon et al. 1996). Any study of a 'community' is somewhat problematic, if performed in strict conformity with the meaning of several theoretical terms, such as species, population, etc. A better-defined task is an operational study including fewer criterions, e.g., an assemblage of living particles. Here, just one criterion (whether a particle is living or not) is critical, but even that criterion alone demands a somewhat complicated judgment. Flow cytometry provides a rapid and reliable method to quantify viable cells in a cell suspension (Agustí & Sánchez 2002). Therefore, a transition to the living matter concept becomes real.

Having operational tools to measure cell viability, we can say that the principal part of each ecosystem is its living matter, i.e., a very specific substance (Vernadsky 1978) composed of a multitude of particles (cells) that have several general properties – growth and fission, reproduction, specific ontogeny (development), metabolism, homeostasis, reaction to external stimuli, and more. Some living particles are very small, some (colonies and organisms) are big, but they also consist of small (μm diameter) living cells (Begon et al. 1996). Each such particle has a limited longevity (a time limit till its death); however, some particles produce new particles resembling their kin precursors. Such ephemeral particles form the eternal successions of particles (the core of each ecosystem), coming and going out via an endless process of self-replication and death. We shall call such transitory particles flow-through elements (Kamenir 2011). Cell viability of natural phytoplankton can be checked by a membrane permeability probe (Perfetto et al. 2006). There are already several very efficient tools (e.g., various flow cytometry and molecular biology tools) for sophisticated, rapid and reliable quantification of viable cells (Agustí & Sánchez 2002; Behar et al. 2010). Determination of viability is critical when evaluating the response of cells to toxic and other strong environmental impacts. Now, cell-viability determination helps us to separate the living matter from other material components of an ecosystem.

Even in the smallest ecosystem, one can see different particles. A special kind of science (taxonomy) can provide a list of such subsets of particles – taxonomic units or species. Modern taxonomic systems help us to optimize the LIM partition into operationally constructed sets of operational taxonomical units (OTUs). Alternatively, we can consider all ecosystem species together and divide the huge mob of particles into a set of distribution functions, applying specific quantitative traits (body mass or volume being the most efficient trait recently). Such distribution functions are called size spectra (Sheldon et al. 1972). Size spectra provide a way of efficient mathematical transformations, as numerous traits of the organism body mass can be evaluated with the help of simple regression models - allometries (Peters 1983). Size spectra and allometry provide highly efficient mathematical means for quantitative analysis of the LIM structure and functioning, while the study of selected by the investigators species and species groups produce practically important results of lesser generality.

Fantastic progress of computers and data acquiring means supports now ever growing hope for general rules in community ecology. Community ecology considers a set of species co-occurring at a given time and place. MacArthur suggested that the goal of community ecology (as of any science) is to find general rules (McGill et al. 2006). Numerous studies, emphasizing the functional relationships between quantitative variables, can support the identification of general patterns and, hence, produce means of biological theory and ecological prediction. The frequency distribution of abundances among species takes a special place in ecology, as a set of taxonomic units demonstrates very general patterns known for numerous sciences considering both, extremely abstract and very real objects (including living communities and human society). The Preston histogram strongly resembles the Boltzmann distribution of kinetic energies among gas molecules and the Pareto distribution of incomes among wage earners (Preston 1950; Nekola & Brown 2007).

As a rule, a phytoplankton assemblage includes a number of species of several types. Major groups of phytoplankton have adopted distinct strategies with associated traits. Therefore, we consider a complicated object composed of several components with contrasting ecological strategies. A formal tool (traits, cell size) supports the development of distribution patterns of major functional groups, i.e., size spectra of specific phyla and of the total phytoplankton. Major functional groups are associated with contrasting environmental regimes. Phytoplankton's ecological niches are largely defined by fundamental physiological processes such as growth and acquisition of a handful of resources (Margalef 1978; Li 2002). Highly heterogeneous time-space structure of the environment (the water mass of a lake or sea) is paralleled with the heterogeneous LIM. Such heterogeneity upgrades the problem to understand how the ecosystem works, but the same heterogeneity provides opportunities for scientific-tool development. This understanding becomes more efficient if several types of analytical methods exist. The inter-comparison of investigative results amplifies our capabilities of prognosis if several types of forecasts produce very similar results. By using long-term monitoring of an ecosystem (in this case, Lake Kinneret, Israel) and combining several types of investigative approaches, a more comprehensive analysis of the community change can be achieved through comparison of results based on organism abundance and taxonomic size spectra.

The aim of this study is to combine two major (i.e., trait-based and taxonomic) branches of aquatic ecology in order to upgrade the ecological prediction capabilities.

MATERIALS AND METHODS

Site Description

Lake Kinneret, Israel, situated ca. 210 m below mean sea level at 32°45'N, 35°30'E (Serruya 1978), is a warm, monomictic lake with a surface area of 170 km^2, and annual maximum depth of approximately 40-44 m. The lake is used for recreation and fishery, but is mainly protected as the major national drinking-water 'store' of Israel.

Therefore, the lake water quality is of prime importance. A routine monitoring program has been carried out since 1969. Since then, Kinneret phytoplankton has exhibited conspicuous ecosystem stability that can be seen from its seasonal species succession and biomass dynamics (Berman et al. 1995). However, drastic changes in taxonomic composition and annual biomass dynamics have become evident since 1993-1996 (Zohary 2004).

Phytoplankton Data Acquisition and Processing

As part of the routine long-term monitoring, phytoplankton samples were collected at 2-week intervals from a fixed pelagic station at the deepest part of the lake, from several depths of the water column above the thermocline. Lugol-preserved samples were brought to the lab and prepared for species identification and counting using the inverted microscope technique (Lund et al. 1958).

All phytoplankton species with cells greater than 2 μm were identified and counted according to species. Two species of relatively common small cyanobacteria were included as they form long filaments. The dominant species with variable cell size (e.g., some dinoflagellate and cyanophyta filaments of variable length) were registered and counted according to size categories. A mean biovolume of the individual cell of each species, or a size category for species counted under several size categories, was calculated from linear microscope measurements and the closest geometric shape (Hillebrand et al. 1999). This cell volume was the parameter used for allocating a taxonomic unit to a size class. Sample processing was described in detail in Zohary (2004) and Kamenir et al. (2006).

Using a long-time monitoring database, we selected a 15-year continuous time interval divided into two parts – a stable (1985-1995) period and an extreme (1996-1999) period.

Species-Abundance Distributions

Species-abundance distributions (SAD or Preston histograms) of all OTUs encountered within a community were developed. We applied the annual maximal abundance estimates (Nj, cell per mL) of OTUj in order to amplify sensitivity while looking for opportunistic and other disturbance-sensitive species. The dependence of maximal annual species abundance

(Nj) on its typical cell volume (Vj) has been described via scatter plots and linear regressions. Each annual species-abundance distribution was also plotted as a histogram of species on the y-axis vs. cell abundance using logarithmic x-axis (Preston 1948). The size class corresponds to an octave, i.e., duplication of the abundance (Preston 1948).

Taxonomic Size Spectra

Each size spectrum was built as the discontinuous frequency distribution of the total number of OTUs to size classes. Size classes were created by cell volume logarithm increments of 0.301. The Vxx notation is used throughout this paper for size classes, where xx is the logarithm [$\log_{10}$ (V)] of the class' right boundary (Kamenir et al. 2006). Since our individual taxon was not always strictly a species, we applied the operational taxonomic unit (Sneath & Sokal 1973) methodology.

Frequency-weighted taxonomic size spectra (FTSS) were built for each phylum and the whole phytoplankton assemblage. Each FTSS applies to each OTU a 'weight' – its frequency rate (FRj) – which is the number of water samples where such an OTU was registered. Frequency-rate values were plotted along the Y axis of the FTSS histogram. These values are the total numbers of times all OTUs belonging to a specific size class (*i*) were registered during the year. The 'team score' of each size class (FRRi) integrates the FRj-weights of all OTUs that fall into the size class *i* (FRji):

$$\text{FRRi} = \Sigma \text{ FRji (Kamenir et al. 2008a).}$$

One should normalize FRj (for standardization) of patterns; however, we safeguard absolute estimates to facilitate the interpretation of the methods and results. A separate FTSS was calculated for the whole phytoplankton and for each its major phylum.

The phytoplankton phyla found in the Kinneret are: Cyanophyta, Diatomea, Chlorophyta, Dinoflagellata, Cryptophyta, Prasynophyta, and Haptophyta (cya2, dia3, chl4, din5, cry6, pras7 and hap9, respectively). Using the histogram and crosstab procedures, we calculated FTSS as the size-frequency distribution of all database records, or all records of relevant phylum. The FTyy notation was applied to refer to a particular year (yy), e.g., FT92 was the FTSS for 1992. A set of annual spectra (FTyy) was developed in the current study, each year, for a set of eight major taxonomic groups (phyla) of Lake Kinneret phytoplankton. The 15-year interval was divided into three 5-year periods. Then, for each of those 5-year periods, average values for each size class were calculated from the FTyy sets (Kamenir 2011).

Similarity and Disturbance Estimation

Treating the taxonomic size spectrum (a discontinuous distribution function) as a numerical vector, we can estimate the similarity/dissimilarity measure for each pair of size spectra with the help of hierarchical cluster analysis (Cha& Srihari 2002). We applied here the Pearson correlation (Sneath & Sokal 1973). The estimates were based on between-group linkage.

The 1st line of the proximity matrix, describing the similarity of each year compared with the 1st year of our study (i.e., 1985) described the dynamics of similarity index r(t). Such a dynamic curve was built for each main phylum of the Kinneret phytoplankton.

A modified assemblage disturbance index, based on opportunistic species (Kamenir & Dubinsky 2012), was produced. The disturbance index was calculated as the average value of 5 log-transformed abundance (Nj) leading species (i.e., ranks 1-5). The difference between the index values of the 1st year (1985) and each following year (i.e., first 5 pairs of log-transformed abundances) were compared by paired t-test (Kamenir & Dubinsky 2012). The SPSS program, version 15.0 (SPSS Inc., Chicago, IL 60606, USA), was used for all statistical procedures.

RESULTS AND DISCUSSION

Scatter plots of maximal annual cell abundance of each species (Nj) vs. cell volume (Vj) have shown a very high variability - almost 6 orders of Nj magnitude for the whole phytoplankton assemblage, including 4 orders for 1 size fraction (Figure 1A). Only very few species had logNj > 4, all of them being small species, i.e., logVj < 2. The cell abundance-vs.-cell volume scatter plot shows that the majority of species occupy the center of the phytoplankton cell-volume range. Only several large species (mainly Dinophyta) occupy the right tail (Figure 1A), where the main part of total biomass is concentrated. The opposite side of the size range also contains very few species, including the abundance leading OTU-s. The smallest algae included just 2 phyla (Cyanophyta and Bacillariophyta); one of them, the Cyanophyta, included a few species (Figure 1B), with very low species biomass but record-high species abundance. The Cyanophyta phylum demonstrated a very steep negative trend of Nj-vs.-Vj (Figure1B). Such a strong trend contrasted all other phyla (Kamenir & Dubinsky 2012).

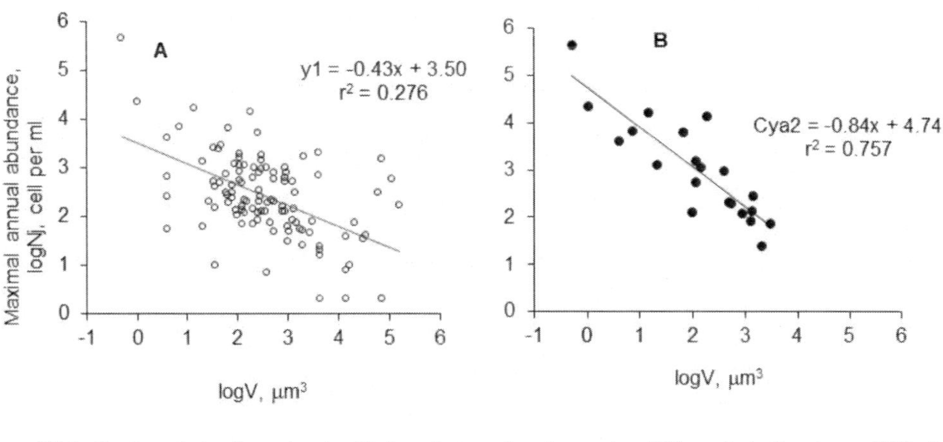

Figure 13-1. Scatter plots of maximal cell abundance of each species (*Nj*) vs. its cell volume (*Vj*), for a disturbed year (1998). A) all phytoplankton species; B) smallest algae - Cyanophyta (cya2). Both axes are log-transformed (after Kamenir & Dubinsky 2012).

Two five-year periods (stable and extreme, 1990-1994 and 1995-1999, A and B, respectively) represent annual histograms of the species number distribution between cell abundance octaves (Figure 2A). The general pattern of histograms was almost the same for the two periods; however, for the extreme period (1995-1999), the bell was broader due to the histogram left-side movement, while the histogram right-hand tail was longer (Kamenir & Dubinsky 2012).

The disturbance index (i.e., the average value of five log-transformed abundance leading species) was produced for each annual dataset (Kamenir & Dubinsky 2012). The index values for 15 consecutive years were compared with that of the first year (1985) of this study (Figure 2B). A drastic change of the index estimates characterized the extreme period of 1995-1999, and also some of the preceding years (1992-1994). The difference between the two of five-year groups (1989-1993 and 1995-1999) was statistically significant (two-sample t-test). The difference between the index values of the year 1985 and almost each year of the 1992-1999 (extreme disturbance) period was also significant (Kamenir & Dubinsky 2012).

While the phytoplankton biomass serves as a traditional criterion for the assemblage disturbance analysis (e.g., bloom of Dinophyta, at the cell volume right tail, $\log V>4$) - the opposite tail, representing small cells ($\log V<1$), deserves attention to determine assemblage disturbances. For several environmental variables (lake water level, its phosphorous load, chloride concentration, and surface layer temperature) statistically significant differences between the annual average estimates of the stable and extreme periods were found. The nutrient stoichiometry and the rising water temperature seem to be plausible factors favoring some Cyanophyta species (Zohary 2004). Results of our study (Figure 2A) support the idea that the length of the histogram right-hand tail can be used as a quantitative disturbance indicator .Species-abundance distribution seems to be a promising tool for the study of fundamental ecological mechanisms. A drastic change in the disturbance index values characterized the extreme period of 1995-1999 (Figure 2B).

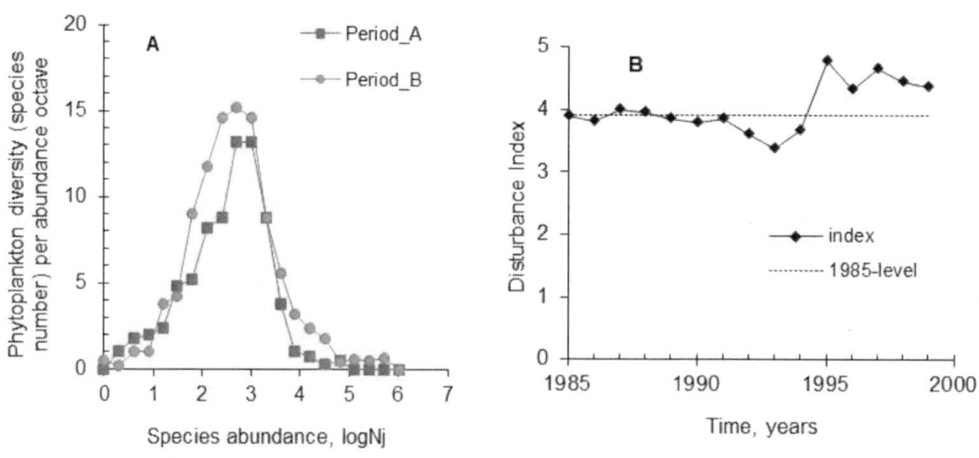

Figure 13-2. Species-abundance distributions of Lake Kinneret phytoplankton. A – Preston histograms of species abundance, representing biodiversity (i.e., species-number) distribution according to octaves of the cell abundance (Nj); periods A and B are 1990-1994 and 1995-1999, respectively. B - dynamics of community disturbance index based on the species abundance dominants, in comparison with the index level for 1985 (after Kamenir & Dubinsky 2012).

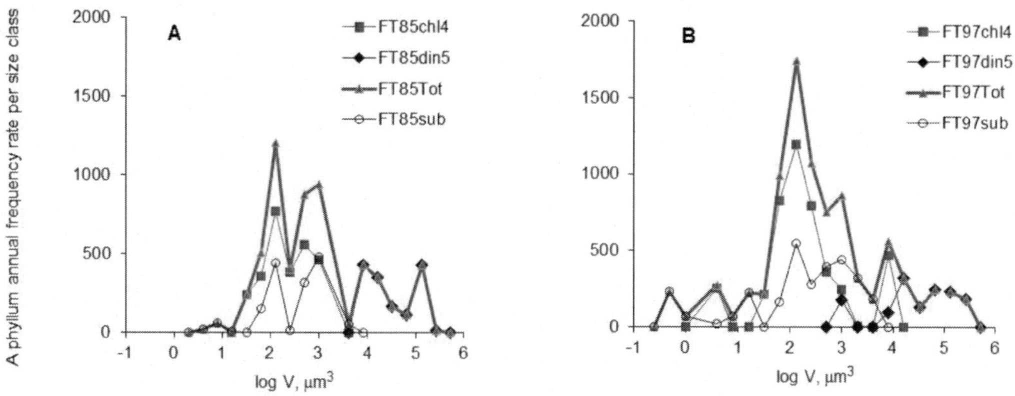

Figure 13-3. The phylum taxonomic structure (FTSS); comparison of three parts of Lake Kinneret phytoplankton. Index "Tot" stands for the sum of all phyla; chl4 and din5 represent two "large" phyla (Chlorophyta and Dinoflagellata), while "sub" stands for the sum of all other phyla (Cyanophyta, Diatomea, Cryptophyta, Prasynophyta, and Haptophyta). Years 1985 and 1997 (plates A and B) represent the stable and extreme periods, respectively. Some phyla exhibit pronounced differences between the two years (after Kamenir 2013).

On the basis of the Preston diagrams, the early-warning information was found for several preceding years (1992-1994), while the causes of the disturbance-index rising variation over a period of several years, before the apparent biomass changes, warrant additional studies. Quantitative similarity comparisons of FTSS demonstrate consistent annual patterns of water mass occupancy by Lake Kinneret phytoplankton assemblage and its main phyla during a long-term period (1985-1994). The extreme period (1995–1999) that followed was characterized by increasing alterations of the annual patterns (Kamenir et al. 2008a; Kamenir 2013). For some phyla, notable changes in the FTSS fine structure and a strong decline in the FTSS year-to-year similarity level were registered during the years 1995-1999. Such spectrum modifications were more pronounced at the phylum level than for the whole phytoplankton assemblage.

A simultaneous comparison of size distribution patterns of the five main phyla (cya2, dia3, chl4, din5, and cry6) demonstrated evident FTSS peak height variability. As a rule, the phylum peaks are separated by a deep trough and are divided by 1-2 orders of cell volume magnitude. The horizontal positions of the main peaks are different for different phyla (Figure 3A and B).

The dendrograms and respective proximity matrices were produced for each of the five main phyla and the whole assemblage. All dendrograms produced very close clustering of the years (Kamenir 2013). The extreme years (1996-1999) were separated from other years (Figure 4). The respective similarity measurements are given in the proximity matrix (Kamenir et al. 2008a). The high similarity described relations during the first 10 years of the phytoplankton assemblage development (Figures 4 and 5), while much lower similarities characterized the extreme years. At the same time, specific phyla often indicated low similarities ($r < 0.5$), which never took place for the whole assemblage. The most interesting phylum of Lake Kinneret was cya2 (Figure 5). Its pronounced deformations during the last years of the studied period produced very low Pearson r estimates.

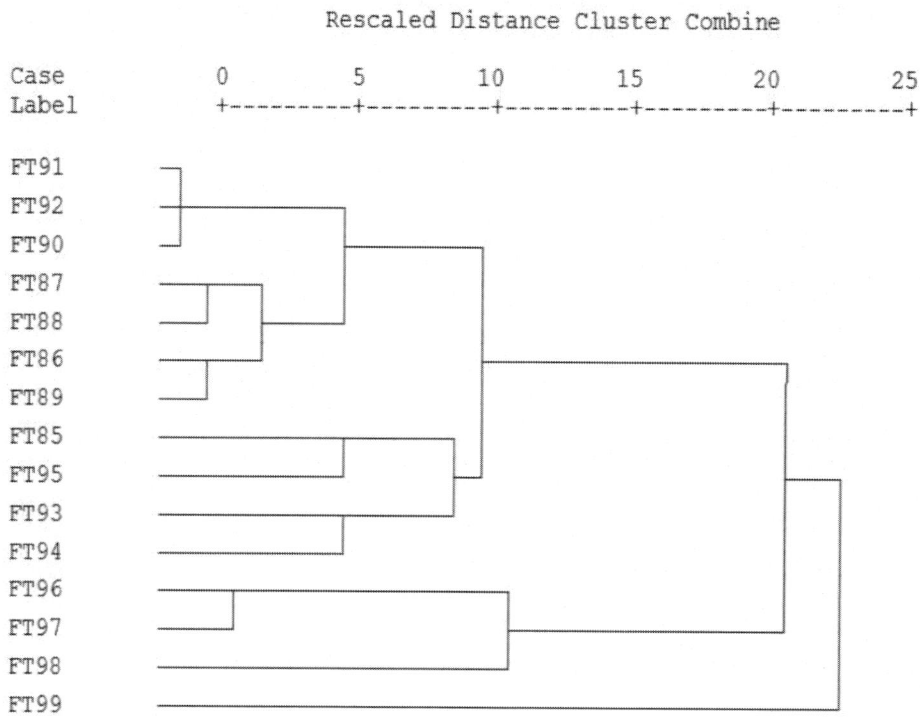

Figure 13-4. The hierarchical cluster analysis dendrogram demonstrating taxonomic structure (FTSS) dynamics of one phytoplankton phyla (Cyanophyta, cya2). FTyy means FTSS for the year yy, 1985-1999 time-range (after Kamenir 2013).

The full r-range of cya2 was 5 times broader than the r-range of the whole phytoplankton. This property may be diagnostically valuable for water-quality maintenance and ecological monitoring. All phyla appear to be 'tied together' by the whole phytoplankton size-distribution limits. The lowest similarity estimates reflected strong FTSS-pattern alterations during the extreme years, such as apparent peak proportion changes and peak side shifts (Figure 3A and B). The results of cluster analysis produced almost identical dendrograms (e.g., Figure 4). The two periods form separate sub-clusters. For the stable period, separate sub-clusters are formed. For the extreme period, one can note apart two pairs, 1996–1997 and 1998–1999, i.e., the suppressed and strong bloom years (Figure 4).

Comparison of all main phylum similarity (r) dynamics showed almost the same general pattern of long-time alterations (Figure 5). The most critical changes - i.e., the lowest r(t) during the years 1994-1995 coincide, but the amplitude of the r-degradation during the years 1994-1995 was different for different phyla (Figure 5). The most sensitive phylum was cya2; its similarity decline was some 3 times more than the r-drop of dyn5 (the phylum dominating the phytoplankton biomass). Therefore, we see a very good concordance of such critical years, although, the respective dynamic curves (Figure 2B and Figure 5) were produced by two very different techniques (cell-abundance histograms and taxonomic structure similarity analysis).

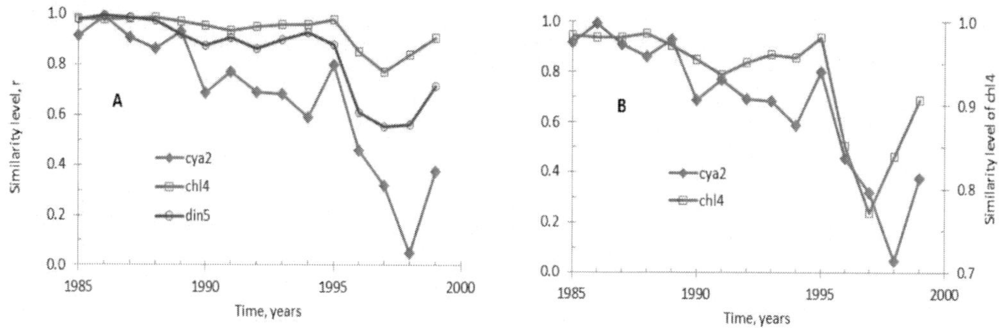

Figure 13-5. Long-time dynamics of taxonomical size structure (FTSS) similarity index (r); comparison of several major taxonomic groups of Lake Kinneret phytoplankton. Cya2, chl4, and din5 mean Cyanophyta, Chlorophyta, and Dinoflagellata, respectively. Note similarity of dynamics of those phyla till 1993, and much more pronounced differences afterwards. The general dynamics pattern is almost the same; plate B demonstrates chl4 curve shape better using the 2nd vertical axis.

CONCLUSION

The whole-assemblage SAD and FTSS demonstrated a high level of similarity. The distribution-pattern changes became evident only during periods of extreme changes of the phytoplankton assemblage annual succession. At the same time, the species-list variation was notable even during the stable period. In such a system, the phylum appears to be at an intermediate level of optimal sensitivity, suitable for the aims of assemblage structural-similarity estimation. Some phyla (Cyanophyta, in our case) demonstrated especially high sensitivity at Lake Kinneret. The SAD and FTSS can be helpful in measuring phytoplankton-structure disturbances. The SAD and FTSS long-term consistency seems to be a general phenomenon produced by mechanisms acting at the ecosystem level of self-organization. A more comprehensive analysis of community change was achieved through comparison of results based on organism abundance and taxonomic size spectra. Comparison of several taxonomic units, seen from two different points of view, supports the identification of general patterns of community structure and dynamics. Such pattern generalization can support the progress of predictive ecology.

Chapter 14

HIGH ORDERLINESS OF PHYTOPLANKTON TAXONOMIC STRUCTURE

INTRODUCTION

Numerous attempts to describe a complex natural community by a single attribute, such as biomass, species richness, equitability, etc., seem to be unrealistic and give way to more detailed and complete pictures (Begon et al. 1996). One of the most important trends in recent time consists of the application of statistical distributions, including both 'ataxonomic' biomass size spectra and distributions of species. Rank-abundance diagrams are very common and have many forms (Whittaker 1965, Brown et al. 2001, Thibault et al. 2004).

Natural communities can be seen as mixtures of species which are unequally successful (Whittaker 1965). If the species are arranged in a sequence ranging from the most to the least important (according to their number of individuals, biomass, etc.) they form a progression with very informative shape and meaning. While such dominance-diversity curves represent a range of intergrading types, they are less orderly and consistent than ecologists would like to see. Nevertheless, this lack of orderliness does not reduce their significance or the generation of ideas about community organization that they offer (Whittaker 1965). Most often resource division among species has been studied using the rank-abundance distributions (RAD) which represent the relative abundances of the species. This distribution of abundance is assumed to reflect the underlying pattern of resource allocation and to provide insight into the processes influencing coexistence. Such analyses typically attempt to determine the mechanisms of the community assembly via two methods, specifically, statistical descriptions and theoretical derivation of the observed distributions (Whittaker 1965, Thibault et al. 2004). While species composition is often highly dynamic in natural communities, some of their features (e.g., total abundance, biomass, and energy flux) are resistant to environmental impacts. The RAD-shape is often discussed as one such resistant property of community structure. Detailed long-term studies show evidence of directional shifts in RAD patterns. Such shifts can support community adaptation, as they alter the resources available to each species, its relative energy use, abundance, and even its presence or absence (Thibault et al. 2004). Unfortunately, the dynamics of RAD is very seldom assessed quantitatively. Such evaluations would allow testing the idea that a consistent pattern of resource allocation (a general division rule) prevails in a natural community. This division rule has been proposed to explain the stability of species richness through time or in response to environmental disturbances (Brown et al. 2001, Thibault et al. 2004).

Progress in the search for such division rules has been limited because of problems of interpretation and the practical difficulty of testing for the best fit (Begon et al. 1996). Thanks to the tremendous progress of computer-based statistical analysis, some very broad generalizations, such as power-law distributions, are receiving ever-growing attention in many fields of modern science (Caldarelli 2007). A very important benefit of such models is their huge commonness providing broad support and experience interchange. Another advantage is the development of theoretical explanations of their origin and self-maintenance (Zipf 1965). Methods of statistical linguistics have recently been applied in various fields of biological science (Caldarelli 2007), for example, they are used to study DNA sequences. The idea is to identify features in the sequences and correlate them with known biological functions (Som et al. 2001). A power-law distribution can be very interesting for aquatic ecology as it resembles the very common power-law approximations of the 'ataxonomic' normalized biomass size spectrum (NBS) (Kerr 1974, Sprules & Munawar 1986), while considering just the opposite (i.e., taxonomic) aspects of the community structure.

Application of a Zipf's-type model could provide a very simple mathematical approximation suitable for the purposes of ecological management and forecast. The same model can be helpful in theoretical studies by providing support to very strong and general criteria for the selection of ecological object-shaping mechanisms.

While considering an ecosystem as a self-maintaining, highly branched feedback loop produced by the closed recycling web (Kamenir 1986; Kamenir & Shteinman 1999), we turn to the dissipative-structure science as the leading methodology needed for selecting the investigative methods and the optimization criteria. Energy dissipation seems to be the main and universal limiting factor in the closed recycling web (Kamenir & Shteinman 1999). Therefore, the least-effort principle (Zipf 1965) can be implied and the distribution laws known from statistical linguistics looked for.

While the rank-abundance models consider mostly the organism number per unit volume or surface (Begon et al. 1996, Thibault et al. 2004), phytoplankton cell abundance seems to be a very flexible, dynamic and heterogeneous property (Heerkloss & Klinkenberg 1998). It is well known that the two most-often-applied descriptors (organism number and its biomass) are misleading when applied to the broad body size intervals of organisms. While biomass criterion overemphasizes the importance of the large organisms, their abundances overemphasize the role of the small ones (Odum 1971). Therefore, it is desirable to exclude organism abundance and its biomass (i.e., organism abundance multiplied by its body mass) from the descriptors applied, since such descriptors can mask more stable patterns to be sought and applied in the forecast (Kamenir 2007). Replacing the abundance figures with their dichotomy (present/absent) leads to incidence-based comparisons (Chao et al. 2005). While considering each species or other taxonomic unit as a specific item of information (a genome, a word-like sequence of several chemical components in a molecule of DNA), we can consider the ecosystem-monitoring dataset as large-scale text. Then, applying statistical models common in modern linguistics and computer sciences, we can proceed from organism numbers to the species-occurrence frequencies (McGeoch & Gaston 2002) reflected in the ecosystem-monitoring database.

Last but not least is the distribution-plot ability to encompass the whole community or assemblage and to demonstrate some reliable typical patterns (Kamenir 2008b). While the species number in natural ecosystems is often measured in the hundreds and thousands, many of them are overlooked or jammed by many kinds of models into artificial blocks. Traditionally, even the rank distributions often consider only one or two dozens of species (e.g., Begon et al. 1996, Thibault et al. 2004). Our study (Kamenir 2011) was aimed at analysis of the whole phytoplankton assemblage in a long time interval.

The aim of this study was a quantitative assessment of the rank-frequency distribution patterns produced by taxonomic units of phytoplankton in Lake Kinneret (Israel) during two contrasting periods, specifically, during years of stability and years of pronounced disturbances.

The working hypothesis was that some common patterns of resource-sharing by taxonomic components of phytoplankton can emerge. Taxonomic structure analysis, exploiting data on water-body occupancy by numerous taxonomic units, can describe structural aspects complementary to more traditional kinds of analyses. The power law can be expected as one of the patterns most common in the numerous fields of science dealing with dissipative structures.

MATERIALS AND METHODS

The site description and data acquisition and processing (including the size spectra development) were explained above (Chapters 9-12). We applied also a rank-frequency model, in which the rank of each species reflected its share in the total dataset, and each unit of information (a database record here) described one species presence in one water sample. The frequency rate (FRj) was estimated for each operational taxonomic unit (OTUj) as the number of samples which contained that OTUj. The total number of OTUs registered in the lake during the eight years described above reached 185. Eight annual datasets, one for each complete annual period, were built (Kamenir et al. 2006) using the same method. Each OTUj list was sorted according to FRj descending order. Thus, each annual graph was created as the rank-frequency distribution of all OTUj registered during one year. The same procedure was done for the pooled dataset which included data for all eight years. A logarithmic transformation - $\log_{10}$ (Y) - was applied to FRj and rank values, producing LogFRj and LogRank estimates, respectively. The curve-fitting procedure was applied to estimate parameters of linear approximation. Analogous procedures were also applied to the abundance, biomass, and typical cell volume of each OTUj to obtain the rank-abundance, rank-biomass and rank-size distributions. While the abundance and biomass of each species showed a broad-scale variation from sample to sample, the maximal estimates of each species were chosen for annual and multi-annual comparisons. The average cell abundance values were estimated from monthly pooled datasets. The histogram procedure was applied to build size-frequency distributions of cell volume (V) and species FRj log-transformed estimates. The size class corresponded to the respective estimate (V or FRj) duplication. Two-tailed two-sample t-tests were used to compare the approximation slope values between the groups of stable and extreme years. A two-tailed paired t-test was applied to compare the *r*-values of the linear regression determination coefficient between the Zipf' s and the log-linear models. The Fisher's Z-transformation was applied to the *r*-values before comparison. The statistical package SPSS 14.0 (SPSS Inc., Chicago, IL) was used for all statistical computations.

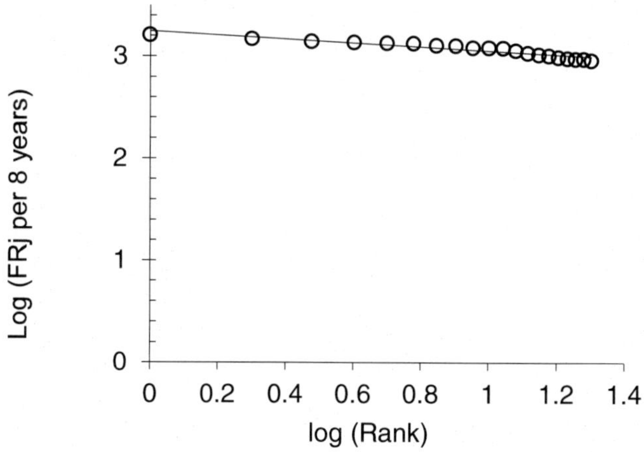

Figure 14-1. A Zipf's (i.e., a power-law) model of the rank-frequency distribution of the Lake Kinneret phytoplankton taxonomic structure for the first 20 species. Log (FRj) = 3.247 - 0.193 log (rank); r^2 = 0.883; $n = 20$; $p < 0.001$. The frequency rate (FRj) of each species represents the number of samples where this species was found during the eight years of monitoring. All phytoplankton species with cell diameter ≥ 2 μm were studied.

RESULTS

A Zipf's (i.e., power-law) model provides a rather good approximation (coefficient of determination of its linear regression $r^2 > 0.88$) of the first 20 ranks (Figure 1). A slight change of the slope can be assumed at both its left and its right extremities. This change becomes obvious when considering a much larger number (n = 100) of ranks (Figure 2). This property of the Zipf's model application is known and has been widely discussed (Som et al. 2001, Caldarelli 2007). A better fit was obtained using a log-linear model (Figure 3).

This linearity seems especially interesting when compared with more traditional (Whittaker 1965, Begon et al. 1996) rank-abundance models. These curves have a less orderly shape that varies from sample to sample (Figure 4).

Each curve can be divided into two-three regions with clearly different slopes. The length of each region and its slope is visibly different each time. More orderly patterns are demonstrated by rank-abundance distributions based on monthly (Figure 4B) and annual datasets (Figure 5A). Rather similar forms for the first 40-50 ranks and much more pronounced variability of the following tail (some 20-30% of taxonomic units) are notable for annual RAD curves (Figure 5A).

This division to the repeatable beginning and a more variable tail also characterizes three other types of rank distributions (Figure 5B-D). For each distribution (Figure 5A-D), the tail (ranks >40-50) has a varying length (i.e., species number), slope and linearity each year. While a pronounced slope change within the first 10-15 ranks is clearly seen for the biomass and cell volume distributions (Figure 5 B and C), the frequency distribution has a smooth beginning part up to ranks 40-50 (Figure 5D). This smooth part is broader when the species number grows, e.g., for long-term datasets (Figures 3 to 5).

The frequency-rate (FRj) logarithm histograms, produced from annual and multi-annual pooled datasets of Lake Kinneret phytoplankton, had slightly different shapes. While the linear FRj-scale produced a highly asymmetric distribution with a very long right tail, the log-transformed estimates (logFRj) showed a steep right slope and a light leftward skewness. The long-term (8-year pooled) histogram was especially close to a lognormal distribution.

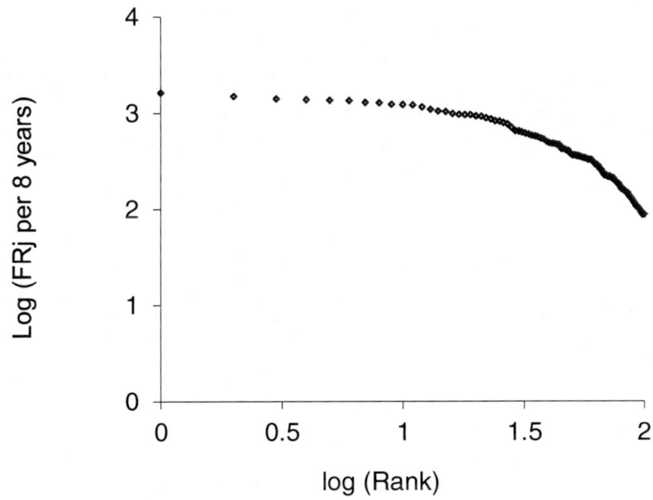

Figure 14-2. A log-log rank-frequency distribution of the Lake Kinneret phytoplankton taxonomic structure for the first 100 species. Note the change of the curvature in the right part of the graph.

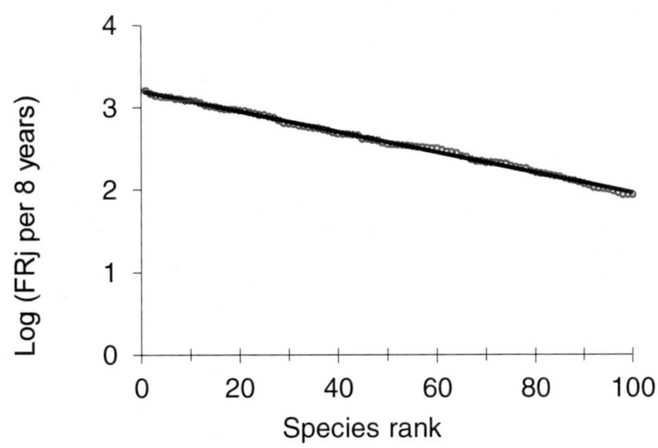

Figure 14-3. A log-linear approximation of Lake Kinneret phytoplankton taxonomic rank- frequency distribution of the first 100 species. Log (FRj) = 3.200 - 0.012 rank; $r^2 = 0.997$; $n = 100$; $p < 0.001$. Note the very high coefficient of determination (r^2) of the linear regression.

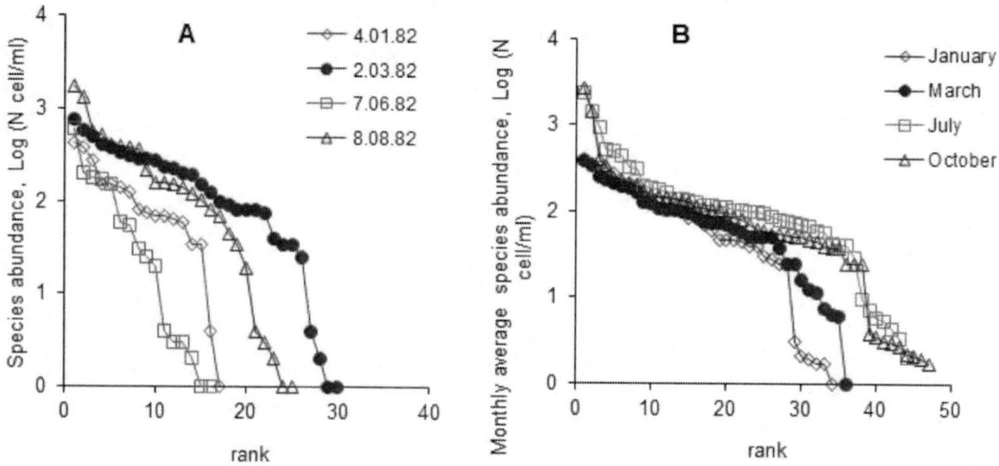

Figure 14-4. Rank-abundance distribution curves of the Lake Kinneret surface-layer phytoplankton for specific water samples (A) and for monthly pooled data (B), year 1982.

While the species number registered during one year was >60 (Kamenir et al. 2006), we have selected the first 40 ranks as the 'beginning part.' The curve fitting (Table 1) shows that the slopes (b) of this part of the rank-frequency distributions do change, but a narrow (some two-times) interval encompasses all these changes. The regression coefficients of determination were higher for the log-linear model (Table 1). Though the slope changes were small (Table 1), all respective Standard Deviations (SDs) were very low, and significant differences were found between the slope values of the stable (1982-1985) and the extreme (1996-1999) years (two-tailed two-sample t-tests) for both the Zipf's ($p = 0.015$) and log-linear ($p = 0.008$) models. The r-vales of the linear regression determination coefficient were significantly higher for the log-linear models than for the Zipf's models (a two-tailed paired t-test; $n = 8$, $p < 0.001$).

Discussion

Consistent Patterns. The main result of this study is evidence of a very high level of orderliness of the natural aquatic assemblage. The existence of ubiquitous statistical distributions of species in natural communities has already emerged in numerous empirical studies and has produced very fruitful theoretical discussions leading to elaborate mathematical models and operational definitions of terms (Preston 1962, Roy et al. 2000, Brown et al. 2001). While such studies consider, as a rule, animals, and mostly terrestrial ecosystems, aquatic ecosystems may also have ubiquitous distributions of this kind, and deserve no less attention. Some aquatic organisms, such as phytoplankton, can be especially valuable for such studies due to the existence of very efficient automated information-acquirement tools. Thanks to the monitoring of numerous water bodies (e.g., lakes and reservoirs), vast databases already exist. They can be reprocessed at rather low costs and lead further to sophisticated mathematical analyses. While zoological studies consider mostly huge time-space scales, the invariant statistical patterns can be found for much smaller objects and periods. Thus, such patterns can be used for the purpose of ecological forecast and monitoring (Kamenir et al. 2006; Kamenir 2008b).

Species rank distributions have been studied for several decades (Whittaker 1965, Begon et al. 1996, Thibault et al. 2004). Forty years ago, R. Whittaker (1965) wrote that these complex and often obscure relationships did not yet lead to an identical formulation. Our study demonstrates rather uniform and consistent patterns found via long-term water-body monitoring, while also showing some factors masking those uniform patterns. The pattern shape and its approximation-model quality depend strongly on monitoring-period length and the number of species being analyzed. Therefore, they depend on the study scale, monitoring team, and resources. While the shape of the curve is conserved from year to year, the species list and their ranks vary. This property can be interpreted as a community adaptation mechanism, which maintains one property (the assemblage rank distribution) while changing another, i.e., rearranging species composition and dominance. The distribution shape may be supported by very general statistical mechanisms. At the same time, its survival can be interpreted as a definite success in community self-maintenance under both stable and disturbed conditions. During the disturbance years (1996-1999) the species number went up and the distribution slope changed while safeguarding the same general pattern (Table 1).

The rank-frequency curves had especially short tail in the left part and especially long main linear region (Figure 5A-D). The longer the species list, the longer is this linear region (Figures 3 to 5). It is especially short for small datasets, like those obtained from one sample (Figure 4A).

While the total number of species (i.e., their maximal rank) and their activity (the maximal FRj values) have increased under ecological-disturbance conditions (years 1996-1999), the linearity of the beginning-part of all curves has been high. Therefore, slope-estimation precision was high, which produced simple quantitative measures to analyze aquatic assemblage structural change (Table 1). These measures (slope and r^2) strongly resemble very common recent NBS approximations (Kerr 1974; Sprules & Munawar 1986), though they are applied not to ataxonomic spectra, but to the assemblage taxonomic composition.

Table 1. Parameters of linear regression approximations*

Year	Zipf-approximation parameters			Log-linear approximation parameters		
	Slope	Slope-SD	r^2	Slope	Slope-SD	r^2
1982	-0.519	0.064	0.636	-0.020	0.001	0.918
1983	-0.441	0.046	0.707	-0.017	0.001	0.947
1984	-0.428	0.047	0.684	-0.016	0.001	0.952
1985	-0.433	0.054	0.631	-0.017	0.001	0.889
1996	-0.372	0.036	0.740	-0.014	0.0005	0.967
1997	-0.311	0.031	0.728	-0.012	0.0004	0.964
1998	-0.382	0.033	0.783	-0.014	0.0005	0.970
1999	-0.259	0.024	0.753	-0.010	0.0003	0.974

*Note. Species rank-frequency distributions were based on the first 40 species; the Zipf's and log-linear models

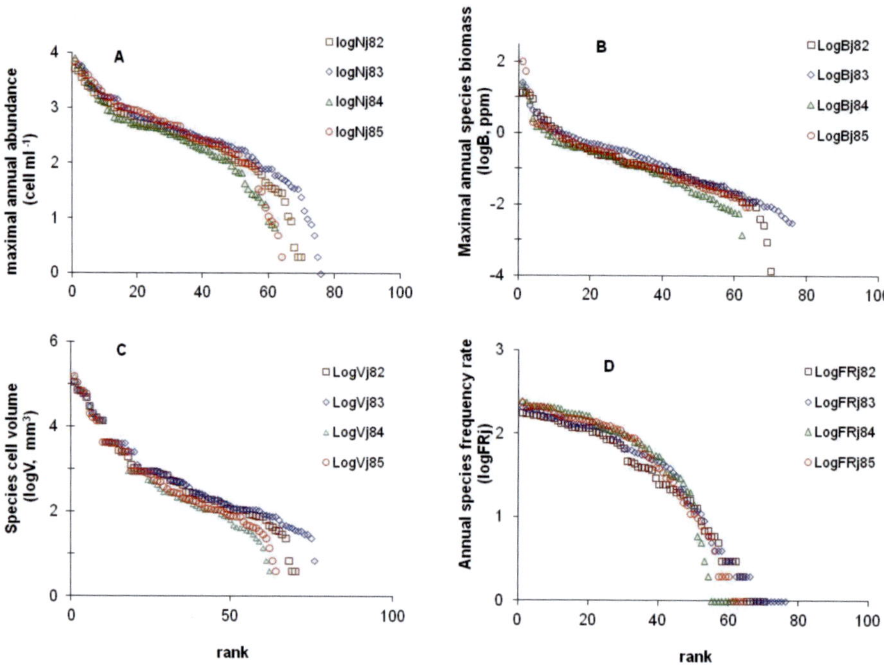

Figure 14-5. Four types of rank distributions of Lake Kinneret phytoplankton taxonomic units. Maximal abundance (Nj, cell ml^{-1}), maximal biomass (Bj, ppm), typical cell volume (Vj, μm^3) and annual frequency rate (FRj) registered for each species during each of four annual periods (1982-1985), A-D, respectively. Logarithmic values were applied for axes Y of all four panels.

An important property of community structure, as reflected by the frequency-rank distributions, is a number of 'rare' species which have FRj estimates some two orders lower than the dominant species FRj (Figure 5D). The rare-species misrepresentation is widely discussed as being very important in many other methods of community structure analysis, including the most traditional ones (Chao et al. 2005). An advantage of the linear models discussed here is that an approximation, excluding the rare-species region (Figure 3), can be very informative, as based on a high species number (n = 100 here) and ignoring only a very small part of the monitoring dataset (some 2% of all registration records). Therefore, it seems reasonable to characterize the curve via application of separate slopes that describe the first and last 10-20% of the rank-distribution. The high- quality fitting of the left part of the curve by the Zipf's and log-linear models supports the application of very simple (linear) regressions. Such simple and precise models can be valuable for the purposes of community-structure modeling and forecast. Several types of rank distributions have this long linear region, if are produced from large datasets (Figures 3 to 5). As they describe different properties of the total assemblage (biomass, abundance, size composition, time-spatial structure), they complement each other and, being applied together, provide a more comprehensive analysis tool (Kamenir 2007). The frequency distribution is especially interesting as being less dependent on several types of errors, including cell counting and cell volume estimation.

While two types of approximation (Zipf's and log-linear) seem to be applicable, scale-free network experience recommends to consider also the log-normal model, as typical (Caldarelli 2007) for the curve types (Figure 5) obtained when all species are taken into account. The lognormal model of the natural assemblage species distribution has been discussed for a long time (Preston 1962, Whittaker 1965). The distributions of FRj come closer to a symmetric bell if a logarithmic transformation is performed. Especially small skewness was observed for the long-term (8-year pooled) histogram. An interesting peculiarity of such histograms is a separate peak in the left region (log FRj < 0.3). This can be interpreted as an artifact produced by the dissection of all low probabilities (i.e., a long left-side tail of the log FRj scale) into 2 estimates of 1 and 0, i.e., to registered and missed species, respectively. An almost symmetric distribution was also evident for species cell-volume logarithm of Lake Kinneret phytoplankton (Kamenir et al. 2006). The difference of each annual size-frequency distribution of species from the lognormal one was non-significant (Kolmogorov-Smirnov test; $p = 0.238$-0.892, $n = 62$-110). The first year (1982) distribution produced the lognormal approximation parameters ($X = 2.72 \pm 0.01$, SD $= 1.01 \pm 0.01$), which suit rather well all other annual distributions, even those obtained 17 years later. This size distribution enhances understanding of the rank-size curves presented above. While the main part of species follow the lognormal distribution (log V from 1.5 to 4.5, i.e., some 50% of the total log V interval), it produces a good linearity of the central region of the rank-size distribution. At the same time, a rather small number of species in the left and right peripheral regions are spread over the other half of the total log V range. They produce much steeper tails on the rank-size curve (Figure 5C). While the Gaussian bell and its deformations explain here the rank-size curves (Figure 5C), the lognormal distribution might explain the other three rank distributions described above (Figure 5). Such analysis gives a sound reason to 'censor' some 10% of the species in each tail region of the rank distributions to obtain more precise slope estimations for the main region (some 80% of species). Then this slope dynamics can be applied as a measure of an integral community structural change. The slope changes seen above (Figures 4 and 5) are caused by the addition of numerous species, mainly within the central cell volume region.

Large-Scale Datasets. Theoretical models and empirical results show that numerous processes can generate very similar-looking distributions that can be separated only on the basis of large, complete datasets of high quality (Caldarelli 2007). Therefore, large-scale studies providing detailed information on the species distributions are necessary. An important distinction of the approach applied here is that we consider OTUs, i.e., groups of cells suitable to be discerned from other such groups (Sneath & Sokal 1973) by computerized tools (Kamenir et al. 2006). OTUj methodology, providing conditions for computer-based methods of particle classification, opens a way to apply automated means such as flow-cytometers and image analyzers. While picoplankton is a very prominent component of phytoplankton assemblages in Lake Kinneret and many other aquatic ecosystems (Malinsky-Rushansky et al. 1995), flow-cytometry should be added to obtain more comprehensive spectra. Thus, such studies can be performed in large aquatic ecosystems within realistic limitations on time intervals, work force, and other resources. The resulting datasets can provide mathematical approximations of high quality necessary to distinguish between the distribution-generating mechanisms. Such mechanism selection is essential for the progress of the aquatic-community self-maintenance theory, which can produce conceptual and mathematical models needed for the purposes of ecological diagnostics and forecast.

Meantime, a plausible explanation can be found in the dissipation-structure theory (Nicolis & Prigogine 1989) and preceding it ideas of G. Cuvier and V. Vernadsky. These concepts have laid the basis for the model we apply, i.e., the ideal minimal ecosystem (IMES) (Kamenir 1986; Kamenir & Shteinman 1999).

Our results demonstrate that not only each organism, but even a population, can be a rather small component of a natural ecosystem, like a sea or lake. Hence, some rules governing the whole living community may be found to be very different from the rules governing a specific population. Consequently, specific properties of the whole living assemblage can exist, suitable for quantitative estimation and forecast, while it is difficult or unreal to forecast the destiny of its numerous specific components.

CONCLUSION

Very simple (linear) models with an extremely small number of parameters that are to be fitted are applicable as good approximations for the annual and multi-annual taxonomic-unit rank-frequency distributions of phytoplankton. Empirical results show a fine tuning of the phytoplankton dominant distribution region (the first 20 ranks) with the help of Zipf's models. A log-linear approximation seems to be preferable to the Zipf's model when considering larger species numbers. Log-linear models also are suitable for the central part of rank-cell volume, rank-abundance and rank-biomass distributions when maximal annual abundance and biomass estimates are chosen for each species. The log-normal distribution produced by the scale-free metabolic web seems to be promising as a possible explanatory model based on concepts of the living whirl (G. Cuvier), living matter (V. Vernadsky), dissipative structures, and biogeochemical cycles. With the help of operational-taxonomic-unit (OTUj) methodology, the large-scale datasets necessary to select the best explanatory models can be obtained via application of automated tools in large aquatic ecosystems.

Chapter 15

A 20TH CENTURY INTERPRETATION

INTRODUCTION

After studying some general concepts of life, biological modeling, and more specific matters concerned with the ecology of aquatic systems, it would be helpful to return to our everyday problems of ordinary life, i.e., life in societies and problems of our environment composed, first of all, of living matter (LIM) and various components of living and non-living matter changed by humans. A vicious circle (or, maybe, just a feedback) characterizes the whole loop, including a person, a society, its environment, and LIM. Such a feedback loop represents a typical structure stabilizing various systems. Feedback structures regulate all components of LIM, including social structures and individuals. Feedback loops hinder our works and intentions but, at the same time, help safeguard many other sweet fruits provided by our Mother-biosphere that support us every day. This situation was described by us a quarter century ago (Kamenir 1991) and continues to be even more interesting now.

Ecology studies cover many exciting local and global problems, including some unpleasant or even dangerous transformations of our environment that took place during the last few centuries, first known as an ecological crisis and later referred to as global change (Vitousek 1994). Currently, the science of ecology is essential for the solution of very acute problems facing mankind. Even the survival of man (i.e., the existence of our species, *Homo sapiens*) is in question nowadays (Sagan 1977). To rescue the health and maintain the survival of many millions of individuals, we must act to prevent undesirable changes in our environment. It is high time to develop a new version of the Hippocratic Oath analogous to the one used by doctors for protecting the health of individuals (Ayzatullin 1989).

The solution of such problems is hindered by several factors, first of all, by their multidisciplinary nature. By and large, attempts of individual specialists to consider such a problem as a whole are considered unrealistic and pretentious; therefore, they are suppressed by scientific ethics. Hence, the drawing up of boundaries defining and describing such interdisciplinary issues is a complicated problem, not to speak of the problem of an operational solution. Therefore, the objective of our study can be the ecological crisis that has been escalating for several centuries (especially during the 20th century and later), considered as a constituent of a complex interaction between man and our milieu, the interaction of human development and adaptation to dynamic environmental conditions prevailing on our planet.

The Time Scale

The time scale of this problem (i.e., the global crisis) should correspond to its spatial scale; hence, it may be several thousand or million years, where a century is a small instant or a dot on a time scale. We consider a special kind of *century description* (specifically, *the 20th century*), a description of the 20th century, made in accordance with some theoretical assumptions developed by previous generations who collected ideas or processed accessible empirical data. Therefore, we have a picture created in the brains of a large set of specialists, a picture somewhat distorted while passing through thick filters of their concepts, ideas, and prejudices.

Several conceptual barriers are concerned with the time-space structure of biosphere, of our LIM. This matter is very special; it is controlled by nonlinear regularities (as a rule, logarithmic or power functions). LIM is quantized, i.e., consists of tremendous quantities of particles (each, in a special closed envelope); hence, each variable describing LIM is limited to integral multiples of individuals (i.e., quants). As a rule, models work well only if there is a multitude of organisms; hence, we can describe properly microorganisms, but not large organisms (including ourselves). For large groups, small groups, and for isolated individuals there are special fields of sciences (e.g., ecology, physiology, medicine).

The image distortion caused by the brain filter seems to be of great importance due to the feedback-loop (i.e., the model – empiric data selection – the model) influence (Prigogine & Stengers 1986), which is often highly underestimated. According to 'the intelligence amplifier scheme,' the generation and selection of models are of fundamental importance. Definition of the problem is of key importance (Shannon 1975). We should clearly define the goals of the study to determine the purpose, what and why we study, and what questions we want to answer. Then, we should determine the boundaries of the entity, restrictions existing in defining the system or process, and investigate how the system works (Shannon 1975).

The model existing in our mind acts as a specific active filter. An autocatalysis loop (i.e., a feedback loop) exists – like in biochemistry or ecology – where a model-predecessor builds its successor with the help of resources (information fragments) sorted out of the previous model's external environment. The initial stage, i.e., the first model selection and the model orientation in an abstract parametric space, modifies all its projections on several drawing planes; therefore, such a projection set is a product of specific selection stages. After such selections and axes rotations of the modeling multidimensional space constructed by us, we obtain submodels of smaller dimensions. Then, we can start a cycle of iterations deriving an ever-growing number of simpler submodels. Some such subsystems are sufficiently small and can suit our analytical capabilities; however, interactions between these numerous submodels become more and more complicated. In this way, we proceed to the problem interpretation, which creates many challenges, but some submodels can be interesting and efficient. The aim of this chapter is such an attempt at problem interpretation.

To further develop a model, we choose some components of our object and its traits that are most important according to the point of view and criteria chosen by us. As cornerstones for our interpretation, we select living matter (Vernadsky 1978) and the theory of dissipative structures (Prigogine & Stengers 1986). The forecasting approach used here is not the calculations of dynamic trajectories (i.e., computer simulations that have become very popular recently), but rather 'the future-in-the-past vision,' i.e., a prognosis obtained as a repetition of an observable quasi-cyclic process, or as a process time-scaled equivalent.

The ideal minimal ecosystem (IMES) model (Kamenir 1986; see Chapter 4 here) is a closed volume (buffer) that maintains a steady state of parameters important for the stability of the closed flow (Ic) of renewal (i.e., the succession of generations – a cyclic process – the birth – development – death – resource release – regeneration – offspring creation) of *flow-through elements* (FELs). The Ic is immersed into the stable medium of the next-level buffer that represents the inner space of the next-level FEL. The most important properties of IMES are its hierarchical structure, cyclic flows and processes, the implementation of each flow by a multitude of FELs, and the equilibrium of several critically important parameters.

The similarity of structures (that follows from the IMES hierarchical structure) suggests that some forms of temporal structures are similar, i.e., a time-scaled equivalence of some processes exists. The most important such resemblance is called the biogenetic law: the development of an embryo (i.e., individual ontogenesis) is a compressed and shortened repetition of taxon development (phylogenesis) (Shklowsky 1980). This law is firmly established, widely discussed, and effectively applied, though numerous problems related to the interpretation of this law exist (Williams & Ebach 2004). Besides numerous, similar, peculiar processes, a set of differences (xenogenesis) exists, i.e., traits of offspring, conflicting with those of either parent. Heterochronies (i.e., changes in the timing of events) are among the most important dissimilarities; there is no stage of complete similarity (Shklowsky 1980). Similarity is observed not only for integral organisms but also for peculiar organs. Such similarity puts the emphasis on analogy within a hierarchically organized system: often, the development of any peculiar organ resembles the evolution of this vertebrata organ (Shklowsky 1980), i.e., a part of an organism. Accordingly with hierarchical model of IMES, the biogenetic law can be interpreted as a flexible process coarsely programmed by blocks (fragments) of hereditary information amalgamating with irregular noises. Often we can recognize a specific block; though, its edges, connections, and place in a chain of blocks can vary. Such programming supports a broad channel of acceptable, asymptotic trajectories.

Another example of analogy is a similarity of processes of succession (i.e., ecosystem ontogenesis) and of evolution (i.e., ontogeny of the highest-level ecosystem – the biosphere). Among the most interesting and important analogous tendencies, one can note: growth of species number during ecosystem succession, progression to better-closed flows (recycling), expansion in space and more far-reaching and all-embracing exploitation of external resources, more intensive resource processing, right-side shift of the SS (i.e., growing role of large structures composing the right tail of distributions), and transition from energy and matter processing to information stores and flows (Khailov 1971; Vernadsky 1978; Odum 1985). The last point also describes a specific phenomenon called cephalization (Cope's law, law of J.D. Dana), from the Greek *κεφαλή* (head), i.e., the concentration of sense organs and information processing at the anterior end of the body, forming a head and brain. Body mass growth involves larger quantities of specific cells and chemical substances used for information accumulation and processing, and the growth of information-processing systems (nerve system, brain, and genome). A similar process might be related to our planetary-scale information handling of the '*noosphere*' (Vernadsky 1978), from the Greek word 'νοῦς' (mind). The concentration of information-processing substances (sense organs, nervous fibers, etc.) takes place during the global evolution from unicellular algae to huge plants and animals, from well-shielded reptiles to naked monkey-type organisms and cephalopods. Such a tendency leads to the formation of a head and brain both during evolution and in an embryo's development.

Noosphere development, which is a global process, was analyzed (Vernadsky 1978) at the beginning of the 20th century. According to Vernadsky, the noosphere is the third phase of development of the Earth, after the geosphere (inanimate matter) and the biosphere (biological life). This new phase has tremendously intensified during several centuries, especially during the last decades, after the fantastic development of electricity, radio, television, computers, and global information nets. Now, satellite-based communications (not to forget the internet and social nets!) have transformed all 'living film' (a thin layer covering the surface of the Earth) into the sphere of ever-growing accumulation and recycling of digital media. This Earth covering includes a global information web and several billion technical apparatuses, realizing various information-processing functions. Noosphere development – a highly complicated growth process – looks just like an 'ordinary catastrophe' (Cobb & Zacks 1985), something resembling exponential growth or S-like curves so common in biology, with precipitously tumbling dynamics. If it does not destroy our planet, this process can take a less-detrimental form, e.g., replacement of the paper-based mass media by TV and other types of on-line streaming, transforming the paper flow into several types of digital electronic streams.

We can describe noosphere development as an ecological succession creating a hierarchical structure that encompasses a huge information recycling web. The typical trends of such a succession include the rightward shift of the SS (including the proliferation of some large structures), brain mass growth in individual ontogenesis (the above-mentioned Dana law), the tremendous biomass growth produced by very large animals with a high proportion of nerve fabric (i.e., *H. sapiens)*. The nerve system evolution leads to advance of various information processing structures (nerve systems) accompanied by accumulation of myelin-type substances (Jahn et al. 2009). Recently (during several millennia, but especially during several decades), information-flow intensification was supported by artificial products, mainly technical tools and technical media (Sagan 1977).

As a component of this process, we see the human population explosion upgrading the global-energy application and anthropogenic information production working together as a positive feedback. This process looks like the biosphere cephalization, including the accumulation of several Mt of human brains (here, Mt is one megaton = one billion men, approximately 1-1.5 kg human brain each). These several Mt of human brains have produced an information explosion during the 20th century and later. This explosion has drastically changed our common way of thinking or paradigm.

Again, by analogy, we can apply the term 'paradigm' (Kuhn 1996), viewed as a distinct set of thought patterns, models of problem formulation, and solution methods dominating in a certain period of time. We can call this object an *everyday paradigm* (EP), meaning a set of behavioral algorithms of moral, ethical, practical, and even esthetical conventions. The EP contains a set of rules created to give solutions to common everyday situations. There are recommendations and attitudes for everyday problems to be formulated and unraveled by everybody, solved not by elaborate calculations but rather as a selection from a bank of traditional recipes. This information flow is enormous and has tremendous importance for our own fate and, maybe, also that of global evolution. Recently, a strong conflict was seen (even in mass media like radio and TV!) between two types of behavior, namely, communism and capitalism and then, of 'European' and 'fundamentalist' (Islamic) ways of thinking. Such large-scale conflicts often lead to religious wars, armed clashes, social unrest, or just discussion on culinary selection.

Up to now, the information cycling flow (Ic) of EP looks like a negligible part of the LIM flow. In comparison with other ecological flows, the EP information flow seems rather small, e.g., look at the world genetic information recycling flow of microorganisms. However, this relatively small flow includes the decision-making process, producing the critically important signals that can break the global-scale equilibrium during a comparatively short time. Several well-known examples of such decisions allow us remember the beginning of the first world war, creation of several huge anthropogenic water bodies (often called seas) along major rivers (Volga, Nile, etc.) and devastation of several huge reservoirs (e.g., the Aral Sea). Much less spectacular but more numerous transformations took place all over the globe during the 20th century, including anthropogenic eutrophication and the biodiversity deterioration that followed. A less spectacular but even more dangerous consequence of this anthropological activity includes the burning of fossil organic matter for the development and maintenance of our artificial infrastructure. One of the consequences of fuel combustion is global warming (Vitousek 1994).

The most active part of our biosphere is living matter. During almost all known history, its small body-size components – bacteria and other unicellular organisms – provide the overwhelming part of the biosphere's total metabolism (matter synthesis and decomposition) (Kamenir & Khailov 1987). Now we see a rapid intensification of anthropogenic information flows that begin to override the natural biosphere flows. First of all, the efficiency of such flows is higher; they produced obvious global changes during fantastically short periods, i.e., decades, compared with millennia of natural global processes. Algorithms applied by science and technology often give efficient solutions to global changes, and can be realized by groups of people inhabiting a small part of Earth. The EP is a critically important factor that changes living-matter properties, organizing mobs into efficient structures capable of selecting the necessary key methods, tuning and upgrading them if needed, to realize the prescriptions directing energy and material flows to a specific goal. One century looks like a very short period that drastically changes our population (*Homo sapiens*) adaptation process, adaptation providing the necessary food, lodging, and adjustment to environmental conditions. At the same time, an ever-growing part of the resources supported by a biosphere is applied to produce, process, and store information. Some such information products favor our population growth and development, but some can damage or destroy all of it.

The EP changes the modes of activity of local and even global populations, just as cells encircled by their common medium change their properties due to hormones excreted by their counterparts, thus changing their common blood. This information medium, integrating data from all interrelated circular flows, envelops every FEL. Each FEL accepts information from an EP, transforming and again releasing it back to the joint EP. The ensemble of closed loops can serve as specific local memory elements, storing parts of EP algorithms. Then, the stored information structure works as a filter for the future subsets of information and can provide some kinds of selective feedback reactions. A multitude of intertwined ringlets create an active medium capable of safeguarding the previous state or – if the negative feedback becomes positive – changing the previous state very quickly.

The eternal cycling can provide stabilization, resembling a gyroscope producing an inertial feedback force. A specific stabilization mechanism can also be created by a pair of flows counterweighing each other. A circular flow can be interpreted as a balanced pair of flows, providing a zero-sum difference, but can become noticeable if one of the two flows has changed 'a little.' Such a relatively small difference (several percent), if it is based on a very high estimate, creates a very strong (and often rapid), unbalanced component. For a system of self-replicating elements, a slight shift of this replication-death balance often means the feedback sign change, i.e., a change between an asymptotic equilibrium and unbalanced exponential growth. The exponential process can produce rapid changes due to a delicate change of a replication coefficient; very often a misbalance produces an S-type curve. Numerous examples of S-type curves were demonstrated during the 20th century: human demography, epidemics, introduction of new species, decline of valuable species, information wars, and digital revolutions. Some resources – energy, carbon dioxide, and numerous industrial and agricultural wastes, demand thorough re-processing, otherwise they can bear catastrophic consequences. Such S-type dynamics can be seen as other examples of standard catastrophes (Cobb & Zacks 1985).

Simulations of mankind dynamics forecasted for the 21st century (Shklowsky 1980) show a collapse after several decades. Even zero-growth population (i.e., strict birth control) is of no help. Population growth is an important component of the ecological crisis, a catalyst of many other processes. There were several conspicuous time scales in this population growth; the most recent one coincided with the science-and-technology revolution during the last 3-4 centuries. The best approximation of this period's demographic dynamics looks like a hyperbole: N(t) = N0/(t0-t), where the critical moment (t0) is 2030±5 yr. Such a formula means that the very law of population increment will radically change soon (Shklowsky 1980).

Such a result can be interpreted according to the model we are applying (IMES). We see not a catastrophe, but reorganization of a biosphere. It may be a reorganization resembling previous ones – analogous to the well-prepared stages of ontogenesis of some other biological systems. Hence, we can 'see the future' if we succeed to find the time-scaling method. Suitable empirical equations (allometries) are known in biology, though their application can be considered as heuristic. The allometry for the intrinsic time scale of an organism is: log(t) ~ b log(W), where b~0.25 (Peters 1983). Assuming that energy dissipation is the main predictor, with respiration (R) allometry, log(R) ~ 0.75 log(W), we can suppose that a time constant τ of a population (consisting of N persons) is: log(τ) ~ 0.25 log(N W).

The *H. sapiens* population estimate (N) produced from global biomass can show 10 g protein m^{-2} multiplied by $5 \cdot 10^{14}$ m^2 (10 g protein ~ 100 g wet weight) ~$5 \cdot 10^{16}$ gww ~ 50 Gt of fresh weight. For a size spectrum of 10 size classes with 100 species at our size class (W ~ 70 kg) (Damuth 1981; Chislenko 1978), we obtain a population estimate of 0.5-1 billion people. Such an estimate is close to demographic estimates of the last millennium (Shklowsky), but has recently grown to some 7 billion. Then, the time-scale ratio gives log(τ) ~ 2-2.5; hence, the time transformation coefficient τ ~ 100-300; therefore, 1 year serves as one 'day' of our population (i.e., the mankind day), and the mankind generation time (τ) of 20 astronomic years transforms into some 4 thousand years.

SEVERAL CENTURIES

The object of our analysis is a 4-century period of demographic explosion and, in parallel, an information explosion. Perhaps the information explosion in Europe (the introduction of book printing to Europe) started a little earlier. One century later, the Protestant Reformation changed work ethics and numerous global processes. Reformation and the Scientific Revolution laid the material basis for the modern, knowledge-based economy. The population-leap period (some 400 years) is comparable with processes seen in ancient Egypt and China (Whittaker & Likens 1975). The 20th century was an important period in the science and technology revolution, which means a qualitative change of the EP.

Nowadays, it is difficult to understand how people lived (just several decades ago) without electricity, telephone, and internet. The scientific revolution of our time was based on many methods of analysis. The analytical approach turned out to be highly efficient in resolving numerous problems of everyday life; hence, we can say the whole EP has changed. The equilibrium of two hemispheres of the human brain is very important (Springer & Deutsch 1998); the same is true for the equilibrium of two hemispheres of the globe. Therefore, the rapid progress of one such hemisphere can break the balance. It seems that the breakdown of EP was already evident at the beginning of the 20th century. Pronounced instability was seen during the 19-20th centuries in a huge space-block of cultures – Balkans-Turkey-Russia-China-Japan. Several years later, all of Europe had fallen and a terrific war expanded worldwide. The first half of the 20th century can be considered among the most dramatic periods in human history.

"God is dead," stated a famous philosopher named Nietzsche in Germany shortly before the 20th century began. Almost the same phrase was formulated by Russian writer D. Merezhkowsky in 1893, i.e., "Never did people feel so strongly in their heart the necessity of faith and understood by reason the impossibility to believe" (Merezhkowsky 1988, p. 47). The last appeals of religious feelings do contradict the last conclusions of modern empirical knowledge (Merezhkowsky 1988). A deep crisis of the main religious dogmas could produce striking changes in the EP.

A description of such processes was given by a famous Russian philosopher, N. Berdyaev, in 1931-1932:

The world has become liquid; there are no solid bodies in it anymore.

Everything is in crisis. There are social and economic crises, but also cultural and spiritual crises. An anti-Renaissance image of a new organism or – more precisely – mechanism emerges that subordinates and absorbs everything.

Foundations of 19th century society should have revealed contradictions and did so, producing disappointment and the reactionary conservatism.

Bitterness of the bifurcated conscience takes the place of man's holistic appreciation, of holism in search of natural shapes.

Self-denial dominates as decomposing, decaying concepts flourish in art. We see everywhere cubistic dismemberment, the decay of man.

The greatest crisis is seen in creative work and culture.

From the depth of our culture emerge some elements of barbarism that inhibit the classical forms of art, science, state, morality, and everyday life.

There are blasts and eruptions emerging from the inside; crises in philosophy and religion. Internal weakness is felt, just as it was at the end of the Ancient World.

All of these points to periodic returns, repetitions, and formal similarities that help us understand our epoch. It resembles the beginning of a new Christian era.

The New History comes to an end and another history begins.

Man becomes a slave of both machines and the material social medium. The social crisis was caused by machinery, but it was mainly a spiritual crisis.

Machinery gives us tremendous forces never seen before; hence, we must take into account that the future of man depends on his spiritual state (Berdyayev 1990, pp. 210-218).

A similar picture was presented in 1920-1924 by another famous thinker, one of the founders of the biosphere and noosphere concepts, V. Vernadsky:

The outstanding process (more powerful and important than all processes that seem important to the contemporaries) is the tremendous spiritual work going everywhere. There are spiritual impressions no less real than material destruction and valuable transpositions. A very deep moral and intellectual crisis involves all thinking mankind. An invisible process – the spiritual result of the world human conscience – is in progress. Its time will come, it will emerge (Vernadsky 1990).

Following the logics of our model (IMES), we can consider such processes as a description of a regular (ontogenetically predetermined) crisis of EP. Three centuries of the analytical-approach advancement, beginning from the Newton laws (Prigogine & Stengers 1968), had led to almost the total dominance of physicalism, and the dominance of classical mechanics (i.e., devices decomposable to parts and deduced through a set of measurable properties). The equilibrium break resulted in a collapse of humanistic renaissance culture. A new stage of succession had begun from the very beginning, i.e., from the fundamentals of morality, spiritualism, and humanity. The EP collapse, i.e., the drastic change in the informational medium, led to social disintegration, EP degradation, and much slower restoration. The closest analogue seems to be a natural community secondary succession or an insect metamorphosis. Its characteristic peculiarities are diminishing homeostasis (Odum 1985; Kamenir 1986), the higher unpredictability of the fate of specific community members, higher dominance of a small number of community members, a decline in ecological diversity (including living creatures, non-living substrates, flows and stores, morphs, and processes), and the diminishing diversity of organism sizes and their life durability. Very characteristic is the change in life strategy, where rapid individual growth dominates instead of a slow elaboration of branching nets of links and adaptations.

The 'social recapitulation' (i.e., a repetition of the societal control-system development) enhances the understanding of some peculiarities of this epoch, including the prominence of specific terms (feudalism, slavery, barbarism), even terms describing ancient tribal systems (e.g., the word *duce*, *fuehrer*). The decline in cultural standards stimulates barbarism in the relations with our dwelling place, in relations with other individuals, peoples, animals, and living beings. Often the *'Hippocratic Oath*' is overridden by the *'Herostratic Oath'* (cf. Herostratus), common with the ancient nobles and despots. There is a tendency to immortalize one's own name at any cost via creation, or even demolition, of something great, like a building, an irrigation system, a city, or a desert.

Chapter 16

RES PUBLICA BIOSPHERE

Nowadays the pendulum swings back. Two great revolutions (in physics and in biology), the development of the general systems theory, the revolution in satellite-based imagery and digital processing, globalization of economics and culture supported by global information and social nets, do manifest the return of holism; however, this ensemble of processes or the EP transformation is far from the equilibrium to be regained. The great forces unleashed by man are still not under our control; some are comparable to natural global processes and endanger our biosphere equilibrium. Urgent actions are needed, but the necessary investments are far smaller than what we have!

Is this really so? The environmental protection programs are not greater than the defense budgets of numerous countries. The 'intra-specific' (i.e., *Homo sapiens*) enmity and our own comfort seem to be more important than the right to existence of all other living beings of Earth. The real problem is our passivity and egoism. The shift is reasonable and moral from the notion of the en-VIRON-ment (from *virun* - circle - a thin belt surrounding us) to the global, integral inhabitance medium.

The notion of environment maintenance, of our common home safety, is important. That is the sole dwelling place of multitudes of living beings that have the same basic rights, including intrinsic rights to life, food, lodging, and protection. Hence, we need a transition from the 'monarchy' (man is 'the king' of Nature) to the approach to oneself as an equal member of *the republic biosphere*. Here, '*Res Publica'* (Lat.) is the common deal, *the common property* existing due to the common, continuous, and correlated activity of many members linked by common aims and interests.

The 'ethnic boundary' (ours/aliens') (Gumilov 1979) should be liquidated, as any member of our biosphere is 'our' (a genuine and lawful participant). The 20th century ecological crisis is just a component of a global crisis – of science, technology, economy, living habits, and logistics – of the all-integrating way of life, i.e., a crisis of the modern paradigm (EP). The tremendously grown capabilities of man (both as an individual and as mankind) should be better controlled; everybody *must be careful* in our fragile world.

The necessary moral standards are known from our earlier joint experience, from ethical teachings of the East and West.

Ethics must relate to everything. Its foundation is our attitude to each other as equals (peers, genuine counterparts), a continuation of the integral whole, according to the ancient principle 'Tat Tvam Asi' (Sanskr.) – 'That is yourself.' Such an approach demands that you see yourself within everything and everything within yourself (Schweitzer 1988). Ethics is the responsibility expanding to infinity, i.e., the unity of each man with everything living. *Ethics is the responsibility.*

REFERENCES

Agustí, S. & Sánchez, M. C. (2002). Cell viability in natural phytoplankton communities quantified by a membrane permeability probe. *Limnology and Oceanography, 47*, 818–828.

Agusty C. S., Duarte, C. M. & Canfield, D. E. (1992). Self-regulation, bottom-up, and top-down control of phytoplankton communities. *Limnology and Oceanography, 37*, 683-687.

Allen C. R., Forys, E. A. & Holling, C. S. (1999). Body mass patterns predict invasions and extinctions in transforming landscapes. *Ecosystems, 2*, 114–121.

Alvarez-Cobelas, M. & Rojo, C. (2000). Ecological goal functions and plankton communities in lakes. *Journal of Plankton Research, 22*, 729–748.

Arin, L., Moran, A. G. & Estrada, M. (2002). Phytoplankton size distribution and growth rates in the Alboran Sea (SW Mediterranean): short term variability related to mesoscale hydrodynamics. *Journal of Plankton Research, 24*, 1019-1033.

Arrhenius, S. (1896). On the Influence of Carbonic Acid in the Air upon the Temperature of the Ground. *Philosophical Magazine and Journal of Science. Series 5*, Volume 41, 237-276.

Ayzatullin, T. A. (1989). Chemistry and chemical industry of the USSR in the structure of the world science and technical progress. *Journal Chemical Society, 1989* (4), 127-138 (in Russian).

Ayzatullin, T. A., Lebedev, V. L. & Khaylov, K. M. (1984). The Ocean: Fronts, Dispersion, Life. Leningrad, Gidrometeoizdat (in Russian).

Barenblatt, G. I. & Monin, A. S. (1983). Similarity principles for biology of the pelagic zone. *Doklady Akad. Nauk SSSR, 269*, 242-244.

Batory, D., Sarvela, J. N. & Rauschmayer, A. (2004). Scaling step-wise refinement. *Software Engineering, IEEE Transactions, 30*, 355-371.

Begon, M., Harper, J. L. & Townsend, C. R. (1996). Ecology: Individuals, populations and communities. 3rd ed. London: Blackwell.

Behar, D., Yunusbayev, B., Metspalu, M., Metspalu, E., Rosset, S., Parik, J., Rootsi, S., Kutuev, I., Yudkovsky, G., Khusnutdinova, E. K., Balanovsky, O., Balaganskaya, O., Semino, O., Pereira, L., Gurwitz, D., Bonne-Tamir, B., Parfitt, T., Hammer, M. F., Skorecki, K. & Villems, R. (2010). The genome-wide structure of the Jewish people. *Nature, 466*, 238–242.

Bell, T. & Kalff, J. (2001). The contribution of picophytoplankton in marine and freshwater systems of different trophic status and depth. *Limnology & Oceanography, 46*, 1243–1248.

Beninca, E., Huisman, J., Heerkloss, R., Johnk, K. D., Branko, P., Van Nes, E. H., Scheffer, M. & Ellner, S. P. (2008). Chaos in a long-term experiment with a plankton community. *Nature, 451*, 822-826.

Benzinger, T. H. (1983). Thermodynamics of Living Matter: Physical Foundations of Biology. *American Journal of Physiology, 244*, R743-50.

Berdyaev, N. A. (1990). Sudba cheloveka v sovremennom mire (The fate of man in modern world). *Novyj Mir (The New World), 1990* (1), 207-232 (in Russian).

Berman, T. & Shteinman, B. (1998). Phytoplankton development and turbulent mixing in Lake Kinneret (1992-1996). *J. Plankt. Res., 20*, 709-726.

Berman, T., Pollingher, U. & Zohary, T. (1998). A Short history of stability and change in phytoplankton populations in Lake Kinneret. *Israeli Journal of Plant Science, 47*, 73-80.

Berman, T., Stone, L., Yacobi, Y. Z., Kaplan, B., Schlichter, M., Nishri, A. & Pollingher, U. (1995). Primary production and phytoplankton in Lake Kinneret: a long-term record (1972-1993). *Limnol. Oceanogr. 40*, 1064-1076.

Bertallanffy, L. (1952). Problems of Life. NY-L: Harpers & Row.

Boudreau, P. R. & Dickie, L. M. (1992). Biomass spectra of aquatic systems in relation to fisheries yield. *Can. J. Fish. aquat. Sci., 49*, 1528-1538.

Brown, J. H. (1995). Macroecology. Chicago, Chicago Univ. Press.

Brown, J. H., Ernest, S. K. M., Parody, J. M. & Haskell, J. P. (2001). Regulation of diversity: maintenance of species richness in changing environments. *Oecologia, 126*, 321-332.

Brown, T. C., Bergstrom, J. C. & Loomis, J. B. (2007). Defining, valuing and providing ecosystem goods and services. *Natural Resources Journal, 47*, 329–376.

Caldarelli, G. (2007). Scale-free networks: Complex webs in nature and technology. Oxford, USA: Oxford Univ Press.

Catling, D. C., Glein, C. R., Zahnle, K. J. & McKay, C. P. (2005). Why O2 Is Required by Complex Life on Habitable Planets and the Concept of Planetary "Oxygenation Time." *Astrobiology, 5*, 415-438.

Cavender-Bares, K. K., Rinaldo, A. & Chisholm, S. W. (2001). Microbial size spectra from natural and nutrient enriched ecosystems. *Limnology & Oceanography, 46*, 778–789.

Cha, S. H. & Srihari, S. N. (2002). On measuring the distance between histograms. *Pattern Recognition, 35*, 1355–1370.

Chao, A., Chazdon, R. L., Colwell, R. K. & Tsung-Jen, S. (2005). A new statistical approach for assessing similarity of species composition with incidence and abundance data. *Ecological Letters, 8*, 148-159.

Charnov, E. L. (1993). Life History Invariants. Oxford, Oxford University Press.

Chisholm, S.W. (1992). Phytoplankton size. In Falkowski, P. G. & A. D. Woodhead (eds.), Primary Productivity and Biogeochemical Cycles in the Sea. N.Y., Plenum Press, 213-237.

Chislenko, L. L. (1981). Struktura fauny i flory v svyazi s razmerami organizmov. Moscow, Moscow Univ. [Structure of fauna and flora as related to the sizes of organisms] (in Russian).

Cobb, L. & Zacks, S. (1985). Applications of Catastrophe Theory for Statistical Modeling in the Biosciences. *Journal of the American Statistical Association, 80*, 793-802.

Cottingham, K. L. & Carpenter, S. R. (1994). Predictive indices of ecosystem resilience in models of north temperate lakes. *Ecology, 75*, 2127-2138.

Cumming, G. S. & Havlicek, T. D. (2002). Evolution, ecology, and multimodal distributions of body size. *Ecosystems, 5*, 705-711.

Cuvier, G. (1827). The animal kingdom arranged in conformity with its organization. London: G. B. Whittaker, vol. 1.

Cyr, H. & Peters, R. H. (1996). Biomass-size spectra and the prediction of fish biomass in lakes. *Canadian Journal of Fisheries & Aquatic Sciences, 53*, 994–1006.

Damuth, J. (1981). Population density and body size in mammals. *Nature, 290*, 699-700.

Dol'nik, V. R. (1978). Energy metabolism and size of animal organisms: Physical foundations of the relationship between them. *Zhurn. obshch. Biologii., 39*, 605—816 (in Russian).

Enquist, B. J. & Niklas, K. J. (2001). Invariant scaling relations across tree dominated communities. *Nature, 410*, 655-660.

Farabee, M. J. (2006). The Online Biology Book. Estrella Mountain Community College, Avondale, Arizona, USA (cited 13.05.15 at http://www2.estrellamountain.edu/faculty/farabee/BIOBK/BioBookTOC.html)

Finn, J. T. (1980). Flow analysis of models of Hubbard Brook Ecosystem. *Ecology, 61*, 562-571.

Firsov, Yu. K. (1979). Poetapnoe opredelenie poverhnosty vodorosley s razlichnym stroeniem sloevishch (Estimation surface of algae with various shapes of tallom). In: Vses. Soveshch. Morsk. Algologyy. *Kiev, Naukova dumka*, 147-149 (in Russian).

Forster, A. C. & Church, G. M. (2006). Towards synthesis of a minimal cell. *Molecular Systems Biology, 2*, paper 45, doi:10.1038/msb4100090.

Fujiki, T. & Taguchi, S. (2002). Variability in chlorophyll a specific absorption coefficient in marine phytoplankton as a function of cell size and irradiance. *Journal of Plankton Research, 24*, 859 874.

Gasol, J. M., del Giorgio, P. A. & Duarte, C. (1997). Biomass distribution in marine planktonic communities. *Limnol. Oceanogr., 42*, 1353-1363.

Gasol, J. M., Guerrero, R. & Pedrós-Alió, C. (1991). Seasonal variation in size structure and prokaryotic dominance in sulfurous Lake Cisó. *Limnology & Oceanography, 36*, 860–872.

Gaston, K. J., Chown, S. L. & Mercer, R. D. (2001). The animal species-body size distribution of Marion Island. *Proc Natl Acad Sci USA, 98*, 14493–14496.

Gilyarov, M. S. (1944). Relationship between size and abundance of animal organisms of the soil. *Dokl. Akad. Nauk SSSR, 43*, 283-285 (in Russian).

Gin, K. Y. H., Chisholm, S. W. & Olson, R. J. (1999). Seasonal and depth variation in microbial size spectra at the Bermuda Atlantic time series station. *Deep-Sea Research I, 46*, 1221-1245.

Glansdorff, P. & Prigogine, I. (1971). Thermodynamic Theory of Structure, Stability and Fluctuations. London, Wiley-Interscience.

Goldman, C. R. (1988). Primary productivity, nutrients, and transparency during the early onset of eutrophication in ultra-oligotrophic Lake Tahoe, California-Nevada. *Limnol. Oceanogr., 33*, 1321-1333.

Goldman, C. R., Jassby, A. & Powell, T. (1989). Interannual fluctuations in primary production: meterological forcing at two subalpine lakes. *Limnol. Oceanogr, 34*, 310-323.

Gorshkov, V. G. (1982). Energetic of Biosphere. Leningrad, LPI (in Russian).

Greenberg, M. J. (1993). Euclidean and non-Euclidean geometries: Development and history. Santa Cruz, University of California.

Grinvald, D. I. & Nikora, V. I. (1988). River Turbulence. Leningrad, Hydrometeoizdat (in Russian).

Gumilyov, L. N. (1979). Ethnogenesis and biosphere of Earth. Leningrad, VINITY (in Russian).

Hambright, K. D., Zohary, T. & Eckert, W. (1997). Potential influence of low water levels on Lake Kinneret: re-appraisal and modification of early hypotheses. *Limnologica, 27*, 149-155.

Harris, D. F. (1908). The Functional Inertia of Living Matter: a Contribution to the Physiological Theory of Life. London, J. & A. Churchill.

Havelicek, T. D. & Carpenter, S. R. (2001). Pelagic species size distributions in lakes: are they discontinuous? *Limnology & Oceanography, 46*, 1021-1033.

Havens, K. E. (2001). Complex analyses of plankton structure and function. *The Scientific World, 1*, 119-132.

Heerkloss, R. & Klinkenberg, G. (1998). A long-term series of a planktonic foodweb: a case of chaotic dynamics. *Verh Internat Verein Limnol, 26*, 1952-1956.

Hemmingsen, A. M. (1960). Energy metabolism as related to body size and respiratory surfaces, and its evolution. *Rep. Steno. Mem. Hospital Nord. Insulinlab., 9*, 1-110.

Hemmingsen, A. M. (1934). A statistical analysis of the differences in body size in related species. *Vidensk. Medd. fra Dansk Natur. Foren., 98*, 125-160.

Higashi, M., Burns, P. B. & Patten, B. C. (1992). Trophic niches and trophic structure of ecosystem: Complementary perspectives through food network unfolding. *J. Theor. Biol., 154*, 57-76.

Hillebrand, H., Durselen, C. D., Kirschtel, D., Pollingher, U. & Zohary, T. (1999). Biovolume calculation for pelagic and benthic microalgae. *Journal of Phycology, 35*, 403–424.

Holling, C. S. (ed.). (1978). Adaptive Environmental Assessment and Management. Chichester: Wiley.

Holling, C. S., (1992). Cross-scale morphology, geometry and dynamics of ecosystems. *Ecological Monographs, 62*, 447-502.

Hoskin, M. (1999). The Cambridge Concise History of Astronomy. Cambridge, Cambridge University Press.

Huisman, J. & Weissing, F. J. (2001). Fundamental unpredictability in multispecies competition. *American Naturalist, 157*, 488–494.

Hunter, D. A., Goldman, C. R. & Bryron, E. R. (1990). Changes in the phytoplankton community structure of Lake Tahoe, California-Nevada. *Verh. Internat. Verein. Limnol., 24*, 505-508.

Hutchinson, G. E. & MacArthur, R. H. (1959). A theoretical ecological model of size distributions among species of animals. *American Naturalist, 93*, 117-125.

Hutchison, C. A., Peterson, S. N., Gill, S. R., Cline, R. T., White, O., Fraser, C. M., Smith, H. O. & Venter, J. C. (1999). Global transposon mutagenesis and a minimal Mycoplasma genome. *Science, 286*, 2165-2169.

Iljichev, V. G. (1982). Stable ecological structures in models with time-lags. *Biophysica, 5*, 858-862 (in Russian).

Irwin, A. J., Finkel, Z. V., Schofield, O. M. E. & Falkowski, P. G. (2006). Scaling-up from nutrient physiology to the size-structure of phytoplankton communities. *J. Plankton Res., 28*, 459-471.

Ivanov, P. P. (1937). General and comparative embryology. Moscow, GIZ (in Russian).

Jahn O., Tenzer S., Werner H.B. 2009. Myelin proteomics: molecular anatomy of an insulating sheath. *Molecular Neurobiology, 40*, 55–72.

Jassby, A. D., Reuter, J. E. & Goldman, C. R. (2003). Determining long-term water quality change in the presence of climate variability: Lake Tahoe (USA). *Can. J. Fish. Aquat. Sci., 60*, 1452-1461.

Jeppesen, E., Lauridsen, T. L., Mitchell, S. F. & Burns, C. W. (1997). Do planktivorous fish structure the zooplankton communities in New Zealand lakes? *New Zealand Journal of Marine Freshwater Research, 31*, 163–173.

John, D. M., Whitton, B. A. & Brook, A. J. (2005). The freshwater algal flora of the British Isles. Cambridge, Cambridge Univ.

Joint, I. (1991). The allometric determination of pelagic production rates. *Journal of Plankton Research, 13*(Suppl.), 69–81.

Jonasz, M. & Fournier, G. (1996). Approximation of the size distribution of marine particles by a sum of log-normal functions. *Limnol. Oceanogr., 41*, 744-754.

Kamenir, Y. (1983). Quantitative Analysis of Internal Fluxes of Ecological Systems. Inst. Biol. South. Seas of Ac. Sci. Ukrainian SSR, Sevastopol, 27pp. Paper deposit at VINITI N2796-V83 (in Russian).

Kamenir, Y. (1986). Size structure of cyclic systems: Parameter interplay. *Ecologiya Morya (Marine Ecology), 24*, 42-51 (in Russian).

Kamenir, Y. (1987). Investigation of Living and Bioinert Matter with Help of Size Spectra. Moscow State University, Moscow, 46pp. Paper deposit at VINITI N5246-V87 (in Russian).

Kamenir, Y. (1991). A XX century crisis interpretation. *Journal Vsesoyuzn. Chem. Obshchestva imeny D. Mendeleev., 36*, 45-56 (in Russian).

Kamenir, Y. (1992). Further evidence of nonnutrient constraints on phytoplankton abundances. *Limnol. Oceanogr., 37*, 681-683.

Kamenir, Y. (2007). Stability of Lake Kinneret phytoplankton structure as evidenced by several types of size spectra. *Fundam Appl Limnol, 168*, 345-354.

Kamenir, Y. (2011a). High Orderliness of Phytoplankton Taxonomic Structure. *International Journal of Medical and Biological Frontiers, 17*, 211-226.

Kamenir, Y. (2011b). Chaos and predictability of lacustrine phytoplankton taxonomic size-structure. (Editor) J A. Daniels, In: Advances in Environmental Research, Vol.15. Nova Science Publishers, 113-134.

Kamenir, Y. (2013). Long-Term Consistency of Main Taxonomic Groups of Lake Kinneret Phytoplankton, pp. 177-198 in: (Editors: Benjamin Veress and Jozsi Szigethy) Horizons in Earth Science Research, v. 9. NY, Nova Science Publishers.

Kamenir, Y. (2015). A Multisided Analysis of Living Matter in an Aquatic Ecosystem. In B. Veress & J Szigethy (Eds.) Horizons in Earth Science Research, v. 13. NY, Nova Science Publishers, 209-219.

Kamenir, Y. & Khailov, K. M. (1987). Metabolic characteristics and total surface area of the living matter of the ocean: A comparison of size spectra. *Oceanology, 27*, 492-496.

Kamenir, Y. & Shteinman, B. (1998). Hydrodynamical approaches for describing the functioning of aquatic ecosystems. In S.S. Wang (ed.) Advances in Hydro-Science and Engineering. Proc. 3rd International Conference on Hydro-Science and Engineering, Cottbus Tech. Univ., Cottbus/Berlin, Germany (CD-ROM) 3, 301-324.

Kamenir, Y. & Shteinman, B. (1999). Multidisciplinary approach to scaling of spectral models of integral aquatic communities. In: Proc 3rd Internatl Symp Ecohydraulics. *Salt Lake City*, pp. 159-181 (CD-ROM).

Kamenir, Y. & Dubinsky, Z. (2012). Disturbance of Opportunistic Small-Celled Phytoplankton in Lake Kinneret. *ISRN Botany, vol.*, 2012, Article ID 123607, 7 pages.

Kamenir, Y., Lopukhin, A. S. & Sysoev, A. A. (1986). Integral Values of ATP of Microplankton of the Black Sea Aquatoria. In: Modern Problems of the Black Sea Oceanology 1, 83-88. Marine Hydrophysical Inst. Ac. Sci. Ukrainian SSR, Sevastopol. Paper deposit at VINITI N1579-V86 (in Rusian).

Kamenir, Y., Walline, P., Shteinman, B., Gophen, M. & Bergstein Ben-Dan, T. (1998). Size structure of aquatic communities: Lake Kinneret case. *Hydrobiologia, 380*, 43-47.

Kamenir, Y., Shteinman, B. & Walline, P. (1999). Influence of basin scale on structure of natural aquatic communities. *Hydrobiologia, 416*, 33-40.

Kamenir, Y., Dubinsky, Z. & Zohary, T. (2004). Size structure stability of phytoplankton in a meso-eutrophic subtropical lake. *Hydrobiologia, 520*, 89-104.

Kamenir, Y., Dubinsky, Z. & Zohary, T. (2006). The long-term patterns of phytoplankton taxonomic size-structure and their sensitivity to perturbation: A Lake Kinneret case study. *Aquatic Sciences, 68*, 490-501.

Kamenir, Y., Dubinsky, Z. & Zohary, T. (2008a). Consistent annual patterns of water mass occupancy are revealed by taxonomic units of Lake Kinneret phytoplankton. *Israel Journal Plant Sciences, 56*, 91-101.

Kamenir, Y., Winder, M., Dubinsky, Z., Zohary, T. & Schladow, G. (2008b). Lake Tahoe vs. Lake Kinneret phytoplankton: comparison of long-term taxonomic size structure consistency. *Aquatic Sciences, 70*, 195-203.

Karmanova, I. V. & Ilina, N. A. (1984). Interrelation between spatial structure and production at several levels of organization (organism, layer, community). *Botanical Journal, 69*, 1593-1602 (in Russian).

Keightley, P. D., Pinharanda, A., Ness, R. W., Simpson, F., Dasmahapatra, K. K., Mallet, J., Davey, J. W. & Jiggins, C. D. (2015). Estimation of the Spontaneous Mutation Rate in Heliconius Melpomene. *Mol. Biol. Evol., 32*, 239–243. doi:10.1093/molbev/msu302.

Kendall, M. A., Warwick, R. M. & Somerfield, P. J. (1997). Species size distributions in Arctic benthic communities. *Polar Biology, 17*, 389-392.

Kerr, S. R. (1974). Theory of size distribution in ecological communities. *J Fish Res Bd Can, 31*, 1859-1862.

Kerr, S. R. & Dickie, L. M. (2001). The biomass spectrum: A predator-prey theory of aquatic production. NY, Columbia Univ. Press.

Khailov, K. M. & Firsov, Yu. K. (1985). Raschet velichiny assimilazionnoj poverhnosty Sargassum mijabei (Sargassacea) na raznyh urovnyah organizazii (Calculation of assimilation surface of Sargassum mijabei (Sargassacea) at various organization levels. *Botan. J., 70*, 215-221 (in Russian).

Khailov, K. M. & Kamenir, Y. (1988). Photoassimilation Surface of Plants and its Correlation with the Respiratory Surface of Animals in Size Rows. *J. Obshch. Biol. (J. Gen. Biol.), 49*, 844-853 (in Russian).

Khailov, K. M. (1994). Vladimir Ivanovivh Vernadsky originator of the biosphere concept. *La Mer, 32*, 1-4.

Khailov, K. M. (1971). Ecological metabolism in the sea. Kiev, Naukova Dumka (in Russian).

Khaylov, K. M. & Parchevskiy, V. P. (1983). Hierarchical Regulation of the Structure and Functions of Marine Plants. Kiev, Naukova Dumka Press (in Russian).

Kiorboe, T. (1993). Turbulence, phytoplankton cell size and the structure of pelagic food webs. *Advances in Marine Biology, 29*, 1-72.

Koga, Y. (2012). Thermal Adaptation of the Archaeal and Bacterial Lipid Membranes. *Archaea*, Article ID 789652, 6 pages. doi:10.1155/2012/789652.

Kovala, P. E. & Larrance, J. P. (1966). Computation of phytoplankton cell numbers, cell volume, cell surface and plasma volume per litre, from microscopical counts. *Department of Oceanography, University of Washington, Seattle, WA, Special Report, 38*, 1-91.

Kozlowski, J. (2002). Theoretical and empirical status of Brown, Marquet and Taper's model of species-size distribution. *Funct. Ecol., 16*, 540-542.

Krebs, H. A. & Kornberg, H. L. (1957). Energy Transformations in Living Matter. Berlin, Springer-Verlag.

Krupatkina, D. K., Lopukhin, A. S. & Kamenir, Y. G. (1987). Size Distribution of Phytoplankton in the Tropical Atlantic and its Effect on the Estimate of Primary Production and Chlorophyll-a in Waters at Different Trophic Levels. *Oceanology, 27*, 348-351.

Kuhn, T. S. (1996). The Structure of Scientific Revolutions, 3rd Ed. Chicago and London: Chicago Press.

Laws, E. A. Redalje, D. G. & Haas, L. W. (1984). High phytoplankton growth and production rate in oligotrophic Hawaiian coastal waters. *Limnol. Oceanogr., 29*, 1161- 1169.

Lebedev, V., Ayzatullin, T. & Khailov, K. (1989). The living ocean. Moscow, Progress.

Levine, I. N. (2002). Physical Chemistry, 5th ed. NY, McGraw-Hill.

Li, W. K. W. & Harrison, W. G. (2001). Chlorophyll, bacteria and picophytoplankton in ecological provinces of the North Atlantic. *Deep-Sea Research II, 48*, 2271–2293.

Li, W. K. W. (2002). Macroecological patterns of phytoplankton in the northwestern North Atlantic Ocean. *Nature, 419*, 154-157.

Litchman, E., Klausmeier, C. A., Schofield, O. M. & Falkowski, P. G. (2007). The role of functional traits and trade-offs in structuring phytoplankton communities: scaling from cellular to ecosystem level. *Ecology Letters, 10*, 1170–1181.

Loeb, J. (1906). The Dynamics of Living Matter. London, Columbia Univ. Press.

Lopukhin, A. S. & Kamenir, Y. G. (1995). Size Spectra of heat production of microplankton of Sevastopol Bay. *Thermochimica Acta, 251*, 53-61.

Lopukhin, A. S., Shaida, V. G. Kamenir, Y. G. & Sysoyev, A. A. (1990). The microplankton of the Tropical Pacific: Distribution and comparative characteristics of size categories as indicated by determinations of ATP and chlorophyll-a. *Oceanology, 30*, 730-739.

Lotka, A. J. (1925). Elements of Physical Biology. Baltimore, Williams & Wilkins.

Lund, J. W. G., Kipling, C. & Le Cren, E. D. (1958). The inverted microscope method of estimating algal numbers and the statistical basis of estimations by counting. *Hydrobiologia, 11*, 143–170.

MacArthur, R. H. & Wilson, E. O. (1967). The theory of island biogeography. Princeton, Princeton Univ. Press.

MacLennan, D. N. & Simmonds, E. J. (1992). Fisheries acoustics. London, Chapman & Hall.

Malinsky-Rushansky, N., Berman, T. & Dubinsky, Z. (1995). Seasonal dynamics of picophytoplankton in Lake Kinneret, Israel. *Freshwater Biol, 34*, 241-254.

Mandelbrot, B. B. (1982). The Fractal Geometry of Nature. San Francisco, W. H. Freeman.

Margalef, R. (1978). Life-forms of phytoplankton as survival alternatives in an unstable environment. *Oceanologica Acta, 1*, 493-509.

Margulis, L. & Rummel, J. D. (2001). Life in NASA and the Rest of the Universe: New Scientific Opportunities. *Bioscience, 51*, 317.

May, R. M. (1981). A cycling index for ecosystems. *Nature, 292*, 105.

McGeoch, M. A. & Gaston, K. J. (2002). Occupancy frequency distributions: patterns, artefacts and mechanisms. *Biol Rev, 77*, 311-331.

McGill, B., Enquist, J., Weiher, E. & Westoby, M. (2006). Rebuilding community ecology from functional traits. *Trends in Ecology & Evolution, 21*, 178-185.

Mendzhul, H. I., Nesterova, N. V., Coryushin, V. A. & Lysenko, T. G. (1985). Tsianofagi - virusy tsianobukteriy (Cyanophages: Viruses of Cyanobacteria). Kiev, Naukova Dumka (in Russian).

Merezhkovsky, D. S. (1988). On the causes of decline and on new trends of modern Russian literature. In: Poetic trends of Russian literature in the end of XIX, beginning of XX century. *Wysshaya Shkola, Moscow (in Russian)*, 46-50.

Milo, R. & Phillips, R. (2015). Cell Biology by the Numbers. NY, Garland.

Moloney, C. L. & Field, J. G. (1989). General allometric equations for rates of nutrient uptake, ingestion, and respirations in plankton organisms. *Limnology and Oceanography, 34*, 1290–1299.

Moore, B. (1921). Biochemistry: A Study of the Origin, Reactions, and Equilibria of Living Matter. London: Arnold.

Morabito, G., Ruggiu, D. & Panzani, P. (2002). Recent dynamics (1995-1999) of the phytoplankton assemblages in Lago Maggiore as a basic tool for defining association patterns in the Italian deep lakes. *J. Limnol., 61*, 129-145.

Morin, A., Bourassa, N. & Cattaneo, A. (2001). Use of size spectra and empirical models to evaluate trophic relationships in streams. *Limnology & Oceanography, 46*, 935–940.

Mouillot, D., Spatharis, S., Reizopoulou, S., Laugier, T., Sabetta, L., Basset, A. & Chi, T. D. (2006). Alternatives to taxonomic-based approaches to assess changes in transitional water communities. Aquatic Conserv.-Mar. and Fresh. *Ecosys., 16*, 469-482.

Muir, B. S. (1969). Gill dimensions as a function of fish size. *Journal Fish Research Board of Canada, 26*, 165-170.

Murray, C. D. (1926). The Physiological Principle of Minimum Work: I. The Vascular System and the Cost of Blood Volume. Proceedings of the National Academy of Sciences of the United States of America, *12*, 207–214.

Nekola, C. J. & Brown, J. H. (2007). The wealth of species: ecological communities, complex systems and the legacy of Frank Preston. *Ecology Letters, 10*, 188–196.

Nicolis, G. & Prigogine, I. (1989). Exploring complexity: An introduction. NY: Freeman.

Nishri, A., Zohary, T., Gophen, M. & Wynne, D. (1998). Lake Kinneret dissolved oxygen regime reflects long term changes in ecosystem functioning. *Biogeochemistry, 42*, 253–283.

Nixon, S. W. (1969). A synthetic microcosm. *Limnol. Oceanogr., 14*, 142-145.

Norussis, M. J. (1998). SPSS 9.0 Guide to Data Analysis. Upper Saddle River (N.J), Prentice Hall.

Odorico, J. S., Kaufman, D. S. & Thomson, J. A. (2001). Multilineage Differentiation from Human Embryonic Stem Cell Lines. *Stem Cells., 19*, 193–204.

Odum, E. P. (1985). Trends expected in stressed ecosystems. *Bioscience, 35*, 419-422.

Odum, E. P. (1971). Fundamentals of ecology. Philadelphia: Saunders.

Pace, N. R. (2001). The universal nature of biochemistry. Proceedings of the National Academy of Sciences of the United States of America, *98*, 805–808.

Patten, B. C. (1985). Energy Cycling in the Ecosystem. *Ecol. Model., 28*, 1-71.

Perfetto, S. P., Chattopadhyay, P. K., Lamoreaux, L., Nguyen, R., Ambrozak, D., Koup, R. A. & Roederer, M. (2006). Amine reactive dyes: an effective tool to discriminate live and dead cells in polychromatic flow cytometry. *Journal of Immunological Methods, 313*, 199-208.

Peters, R. (1986). The role of prediction in limnology. *Limnol. Oceanogr. 31*, 1143-1159.

Peters, R. H. (1983). The ecological implications of body size. Cambridge: Cambridge Univ. Press.

Petersen, F. (1983). Population dynamics and production of Daphnia galeata (Crustacea, Cladocera) in Lake Esrom. *Holarctic Ecol., 6*, 285-294.

Pierce, W. S. & Hendrix D. L. (1979). Utilization of Enzyme Markers to Determine the Location of Plasma Membrane from Pisum Epicotyls on Sucrose Gradients. *Planta, 146*, 161-169.

Platt, T. & Denman, K. (1978). The structure of pelagic ecosystems. *Rapport Proceedings Reunion Conseil International Exploration de la Mer, 173*, 60-65.

Pollingher, U. (1986). Phytoplankton periodicity in a subtropical lake (Lake Kinneret, Israel). *Hydrobiologia, 138*, 127-138.

Pollingher, U. (1991). A long term study of Chroococcus population in Lake Kinneret (Israel). *Algological Studies, 64*, 377-383.

Pollingher, U. & Serruya, C. (1976). Phased division of Peridinium cinctum fa. westii (Dinophyceae) and development of the Lake Kinneret, Israel bloom. *Journal of Phycology, 12*, 162-170.

Pollingher, U., Hadas, O., Yacobi, Y. Z., Zohary, T. & Berman, T. (1998). Aphanizomenon ovalisporum (Forti) in Lake Kinneret (Israel). *Journal of Plankton Research, 20*, 1321-1339.

Preston, F. W. (1948). The commonness, and rarity, of species. *Ecology, 29*, 254-283.

Preston, F. W. (1950). Gas laws and wealth laws. *The Scientific Monthly, 71*, 309– 311.

Preston, F. W. (1962). The canonical distribution of commonness and rarity: Part I. *Ecology, 43*, 185-2 15.

Prigogine, I. & Stengers, I. (1984). Order out of Chaos: Man's New Dialogue with Nature. NY, Bantam.

Quintana, X. D., Comin, F. A. & Moreno-Amich, R. (2002). Biomass-size spectra in aquatic communities in shallow fluctuating Mediterranean salt marshes (Emporda wetlands, NE Spain). *Journal of Plankton Research, 24*, 1149-1161.

Raffaelli, D., Hall, S., Emes, C. & Manly, B. (2000). Constraints on body size distributions: An experimental approach using a small-scale system. *Oecologia, 122*, 389–398.

Rasmussen, S., Chen, L., Deamer, D., Drakauer, D. C., Packard, N. H., Stadler, P. F. & Beadau, M. A. (2004). Evolution: Energy Transformations from Nonliving to Living Matter. *Science, 303*(5560), 963-65.

Raymont, J. E. G. (1980). Plankton and Productivity in the Oceans. Vol. 1: Phytoplankton. Oxford, Pergamon.

Reynolds, C. S. (2002). On the interannual variability in phytoplankton production in freshwaters. In: Williams PJIB, Thomas DN, Reynolds CS (eds). Phytoplankton productivity: Carbon assimilation in marine and freshwater ecosystems. Oxford, Blackwell, 187-221.

Rodriguez, F., Fernandez, E., Head, R. N., Harbour, D. S., Bratbak, G., Heldal, M. & Harris, R. P. (2000). Temporal variability of viruses, bacteria, phytoplankton and zooplankton in the western English Channel off Plymouth. *Journal of Marine Biological Association UK, 80*, 575-586.

Rodriguez, J. & Mullin, M. M. (1986). Relation between biomass and body weight of plankton in a steady state oceanic ecosystem. *Limnology & Oceanography, 31*, 361-370.

Roy, K., Jablonski, D. & Martien, K. K. (2000). Invariant size-frequency distributions along altitudinal gradient in marine bivalves. *Proc Natl Acad Sci USA, 97*, 13150-13155.

Rubner, M. (1883). Uber den Einfluss der Korpergrosse auf Stoff und Kraftwechsel. *Zeitschrift fur Biologie, 19*, 536–562.

Sagan, C. (1977). The Dragons of Eden: Speculations on the Evolution of Human Intelligence. NY, Random House.

San Martin, E., Harris, R. P. & Irigoien, X. (2006). Latitudinal variation in plankton size spectra in the Atlantic Ocean. *Deep-Sea Research II, 53*, 1560-1572.

Schaub, G. & Turek, Th. (2011). Energy Flows, Material Cycles and Global Development. A Process Engineering Approach to the Earth System. Berlin-Heidelberg, Springer, 204pp.

Scheffer, M., Rinaldi, S., Huisman, J. & Weissing, F. J. (2003). Why plankton communities have no equilibrium: solutions to the paradox. *Hydrobiology, 491*, 9–18.

Schmidt-Nielsen, K. (1984). Why is Animal Size So Important? Cambridge: Cambridge University Press.

Schweizer, A. (1988). The weltanschauung of Indian thinkers: mystics and ethics. In West-East. Moscow, Nauka (in Russian), 209-233.

Schwinghamer, P. (1981). Characteristic size distribution of integral benthic communities. *Canadian Journal of Fisheries & Aquatic Sciences, 38*, 1255-1263.

Seager, S. (2013). Exoplanet Habitability. *Science, 340*, 577-581.

Serruya, C. (ed). (1978). The Kinneret. Monographia Biologica. The Hague: Dr W. Junk Publishers.

Shakun, J. D., Peter, U., Clark, Feng He., Shaun, A., Marcott, Alan C., Mix, Zhengyu., Liu, B., Otto-Bliesner, Schmittner A., & Bard, E. (2012). Global warming preceded by increasing carbon dioxide concentrations during the last deglaciation. *Nature, 484*, 49-55.

Shannon, R. E. (1975). Systems Simulation: The Art and Science. Upper Saddle River, Prentice-Hall.

Sheldon, R. W., Prakash, A. & Sutcliffe, W. H. (1972). The size distribution of particles in the ocean. *Limnology and Oceanography, 17*, 327-340.

Shklowsky, I. S. (1980). Universe, life, mind. Moscow, Nauka (in Russian).

Shushkina, E. A., Vinogradov, M. Y. & Vostokov, S. P. (1984). Determination of primary production and heterotropic destruction in the epipelagic zone of the ocean. *Okeanology, 24*, 130-138.

Sieburth, J. M., Smetacek, V. & Lenz, J. (1978). Pelagic ecosystem structure: Heterotrophic compartments and their relationship to plankton size fractions. *Limnol. Oceanogr., 23*, 1256-1263.

Sin, Y., Wetzel, R. L. & Anderson, I. C. (2000). Seasonal variations of size-fractionated phytoplankton along the salinity gradient. *Journal of Plankton Research, 22*, 1945-1960.

Smith, F. A., Brown, J. H., Haskell, J. P., Lyons, S. K., Alroy, J., Charnov, E. L., Dayan, T., Enquist, B. J., Ernest, S. K. M., Hadly, E. A., Jones, K. E., Kaufman, D. M., Marquet, P. A., Maurer, B. A., Niklas, K. J., Porter, W. P., Tiffney, B. & Willig, M. R. (2004). Similarity of mammalian body size across the taxonomic hierarchy and across space and time. *American Naturalist, 163*, 672-691.

Smith E. M. & Kemp W. M. 2001. Size structure and the production/respiration balance in a coastal plankton community Limnology & Oceanography 46, 473–485.

Sneath, P. H. A. & Sokal, R. R. (1973). Numerical Taxonomy. San Francisco, Freeman.

Sober, E. (1983). Parsimony in Systematics: Philosophical Issues. *Annual Review of Ecology and Systematics, 14*, 335–357.

Sokal, R. R. & Rohlf, F. J. (1995). Biometry: The principles and practice of statistics in biological research. San Francisco, Freeman.

Som, A., Chattopadhyay, S., Chakrabarti, J. & Bandyopadhyay, D. (2001). Codon distributions in DNA. *Physical Review E*, 63-051908, 1-8.

Sorokin, Yu. I. (1978). Decomposition of organic matter and nutrient regeneration. In: Kinnne O. (Ed.). *Marine Ecology, Vol., 4*. Dynamics. London, Wiley, pp. 501—616.

Sosik, H. M. & Olson, R. J. (2007). Automated taxonomic classification of phytoplankton sampled with imaging-in-flow cytometry. *Limnol. Oceanography: Methods, 5*, 204–216.

Springer, S. P. & Deutsch, G. (1998). Left Brain, Right Brain: Perspectives on Cognitive Neuroscience, (5th ed.). New York: W.H. Freeman.

Sprules, W. G. (1988). Effects of trophic interactions on the shape of pelagic size spectra. Verh. Internat. *Verein. Limnol., 23*, 234-240.

Sprules, W. G. & Goyke, A. P. (1994). Size-based structure and production in the pelagia of Lakes Ontario and Michigan. *Canadian Journal of Fisheries & Aquatic Sciences, 51*, 2603-2611.

Sprules, W. G. & Munawar, M. (1986). Plankton size spectra in relation to ecosystem productivity, size, and perturbation. *Can J Fish Aquat Sci, 43*, 1789-1794.

Stefan, I. (2007). Distributed Database Design – Top-Down Design. *Acta Electrotechnica, 1*, 56-59.

Steinberg, C. E. W., Shaefer, H., Siedler, M. & Beisker, W. (1996). Ataxonomic assessment of phytoplankton integrity by means of flow cytometry. *Archiv. Toxicol. Suppl., 18*, 417-434.

Stephens, C. A. (1888). Living Matter: Its Cycle of Growth and Decline in Animal Organisms. Norway Lake (Me), Laboratory Co.

Stone, L. & Berman, T. (1993). Lake Kinneret: A feed-back model for carbon flux through planktonic biota. *Limnol. Oceanogr., 38*, 1680-1695.

Strayer, D. L., Power, M. E., Fagan, W. F., Pikett, S. T. A. & Belnap, J. (2003). A Classification of Ecological Boundaries. *Biosciences, 58*, 723-729.

Takahashi, M. & Bienfang, P. K. (1983). Size structure of phytoplankton biomass and photosynthesis in subtropical Hawaiian waters. *Marine Biology, 76*, 203—211.

Taub, F. B. (1969). A biological model of a freshwater community: A gnotobiotic ecosystem. Limnol. *Oceanogr., 14*, 136-142.

Thibault, K. M., White, E. P. & Ernest, S. K. M. (2004). Temporal dynamics in the structure and composition of a desert rodent community. *Ecology, 85*, 2649-2655.

Thiebaux, M. L. & Dickie, L. M. (1993). Structure of the body-size spectrum of the biomass in aquatic ecosystems: a consequence of allometry in predator-prey interactions. *Canadian Journal of Fisheries & Aquatic Sciences, 50*, 1308–1317.

Thingstad, T. F. & Sakshaug, E. (1990). Control of phytoplankton growth in nutrient recycling ecosystems. Theory and terminology. *Marine Ecology Progress Series, 63*, 261-272.

Tipler, P. & Llewellyn, R. (2002). Modern Physics, 4th ed. Bedford, W.H. Freeman.

Tseytlin, V. B. (1986). Energetics of Deep-Water Pelagic Communities. Moscow, Nauka (in Russian).

Turkia, J. & Lepisto, L. (1999). Size variations of planktonic Aulacoseira Thwaites (Diatomae) in water and in sediment from Finnish lakes of varying trophic state. *Journal of Plankton Research, 21*, 757-770.

Utermöhl, H. (1958). Zur Vervollkömmung der quantitativen Phytoplankton-Methodik. *Mitteilungen Internationale Vereinigung fuer Theoretische und Angewandte Limnologie, 9*, 1–38.

Vander Zanden, M. J., Chandra, S., Allen, B. C., Reuter, J. E. & Goldman, C. R. (2003). Historical food web structure and restoration of native aquatic communities in the Lake Tahoe (California-Nevada) basin. *Ecosystems, 6*, 274-288.

Vernadsky, V. I. (1929). La Biosphere. Paris: F. Alcan (in French).

Vernadsky, V. I. (1978). Zhivoe Veshchestvo (Living Matter). Moscow: Nauka (in Russian).

Vernadsky, V. I. (1988). Phylosophical thoughts of a naturalist. Moscow, Nauka (in Russian).

Vernadsky, V. I. (1990). Enough of blood and sufferings. In: Vek XX i Mir (XX Century and Peace). Moscow, Nauka (in Russian), 24-30.

Vernadsky, V. I. (1998). The Biosphere. New York: Springer-Verlag.

Vitousek, P. M. (1994). Beyond Global Warming: Ecology and Global Change. *Ecology, 75*, 1861-1876.

Vitousek, P. M., Mooney, H. A., Lubchenco, J. & Melillo, J. M. (1997). Human domination of earth's ecosystems. *Science, 277*, 494-499.

Walline, P. D., Pisanty, S., Gophen, M. & Berman, T. (1993). The ecosystem of Lake Kinneret, Israel. In V. Christensen & D. Pauly (eds.), Trophic models of aquatic systems. *ICLARM Conf. Proc., 26*, 103-109.

Warwick, R. M. (1984). Species size distributions in marine benthic communities. *Oecologia, 61*, 32-41.

West, G. B. & Brown, J. H. (2005). The origin of allometric scaling laws in biology from genomes to ecosystems: towards a quantitative unifying theory of biological structure and organization. *Journal of Experimental Biology, 208*, 1575-1592.

West, G. B., Brown, J. H. & Enquist, B. J. (1997). A general model for the origin of allometric scaling laws in biology. *Science, 276*, 122–126.

West, G. B., Brown, J. H. & Enquist, B. J. (1999). A general model for the structure and allometry of plant vascular systems. *Nature, 400*, 664-667.

West, G. B., Woodruff, W. H. & Brown, J. H. (2002). Allometric scaling of metabolic rate from molecules and mitochondria to cells and mammals. *Proc. Natl. Acad. Sci. USA, 99*, 2473-2478.

Wetzel, R. G. (2001). Limnology: Lake and River Ecosystems. 3rd Edition. San Diego, Academic Press.

White, C. R. & Seymour, R. S. (2003). Mammalian basal metabolic rate is proportional to body mass2/3. *PNAS, 100*, 4046-4049.

White, E. P., Ernest, S. K. M., Kerkhoff, A. J. & Enquist, B. J. (2007). Relationships between body size and abundance in ecology. *TRENDS in Ecology and Evolution, 22*, 323-330.

White, L., Jr. (1967). The Historical Roots of Our Ecologic Crisis. *Science, 155*, 1203-1207.

Whittaker, R. H. (1961). Experiments with radiophosphorus tracer in aquarium microcosms. *Ecol. Monogr., 31*, 157-188.

Whittaker, R. H. (1965). Dominance and diversity in land plant communities. *Science, 147*, 250-260.

Whittaker, R. H. & Likens, G. E. (1975). The biosphere and man. In: Lieth H. and R.H. Whittaker (Es.). Primary Productivity of the Biosphere. Berlin, Springer-Verlag, pp. 305-328.

Yang, Y., Yokobori, S. & Yamagishi, A. (2009). Assessing Panspermia Hypothesis by Microorganisms Collected from The High Altitude Atmosphere. *Biological Sciences in Space, 23*, 151-163.

Yentsch, C. M., Horan, P. K., Muirhead, K., Dortch, Q., Haugen, E., Legendre, L., Murphy, L. S., Perry, M. J., Phinney, D. A., Pomponi, S. A., Spinrad, R. W., Wood, M., Yentsch, C. S. & Zahuranec, B. J. (1983). Flow cytometry and cell sorting: a technique for analysis and sorting of aquatic particles. *Limnology & Oceanography, 28*, 1275-1280.

Zipf, G. K. (1965). Human behaviour and the principle of least effort. NY, Hafner.

Zohary, T. (2004). Changes to the phytoplankton assemblage of Lake Kinneret after decades of a predictable, repetitive pattern, *Freshwater Biology, 49*, 1355-1371.

INDEX

A

B

C

D

E

F

G

H

I

L

M

N

O

P

R

S

T

U

V

W